Demystifying Switching Power Supplies

Demystifying Switching Power Supplies

Raymond A. Mack, Jr.

AMSTERDAM • BOSTON • HEIDELBERG • LONDON
NEW YORK • OXFORD • PARIS • SAN DIEGO
SAN FRANCISCO • SINGAPORE • SYDNEY • TOKYO

Newnes is an imprint of Elsevier

ELSEVIER

Newnes

Newnes is an imprint of Elsevier
30 Corporate Drive, Suite 400, Burlington, MA 01803, USA
Linacre House, Jordan Hill, Oxford OX2 8DP, UK

Recognizing the importance of preserving what has been written, Elsevier prints its books on acid-free paper whenever possible.

Library of Congress Cataloging-in-Publication Data
Mack, Raymond.
  Demystifying switching power supplies / Raymond Mack.
     p.   cm.
  Includes bibliographical references and index.
  ISBN 0-7506-7445-8 (alk. paper)
  1. Switching circuits—Design and construction. 2. Power semiconductors—Design and construction. 3. Semiconductor switches—Design and construction. 4. Switching power supplies—Design and construction. I. Titile.
  TK7868.S9M24 2005
  621.31'7—dc22   2004029371

British Library Cataloguing-in-Publication Data
A catalogue record for this book is available from the British Library.

For information on all Newnes publications
visit our Web site at www.books.elsevier.com

Transferred to digital printing in 2009.

Contents

Preface

This book is intended for those who need to understand how a switching power supply works. I intend to provide enough information so you can intelligently specify a custom off-line supply from a power supply manufacturer. You should also gain enough information to be able to design a DC–DC converter. I have included basic analog design information for those whose primary electronics background is not analog circuits. Then I build on that basic information to show how to design and analyze practical switching power supplies. Those with a strong background in analog circuitry may want to skim over the preliminary data.

In numerous places I skip over the details of derivations and transformations of equations. The details of those transformations are left as an exercise for the reader.

There are two broad classes of power supplies: linear and switching. Linear supplies use time continuous control of the output. Switching supplies are time-sampled systems that use rectangular samples to control the output. This book explores each of the variations of switching power supplies.

Acknowledgments

Like most work, this book is built on the efforts of many others. I wish to acknowledge the large contribution to my understanding of switching power supplies by the authors of the Motorola application book *Linear/Switchmode Voltage Regulator Handbook*, the International Rectifier *HDB-3 Power MOSFET HEXFET Databook*, and the Philips *Switch Mode Power Supply* Semiconductor application book (an excellent book but available only on their website).

I also wish to acknowledge the gracious contributions by Linear Technology Corporation. Linear Technology gives away their program SwitcherCAD III. It is intended for use by their customers, but it is free to all who want to use it. Most of the schematics in this book were initially prepared using the drafting functions of SwitcherCAD III.

Introduction

The principles of switching power supplies have been used for over 100 years (though people didn't know that's what they were). The ignition system used in a gasoline engine was the earliest version of a flyback switching power supply. The next general use of switching supplies was in the high voltage section of televisions. Again, this is an example of a rudimentary flyback supply. The flyback name comes from the short time period where the spot on the television CRT is moved from the right side of the screen back to the left side of the screen (it would "fly back"). The rapid change in current in the deflection coil causes a very large voltage to be generated. This was used to advantage in televisions to create the large acceleration potential necessary for the CRT.

Widespread switching supply use was limited to television high voltage service until the late 1960s because of limited capabilities of the three major components in a switching supply: the magnetics, the switch, and the rectifier. Components were available for switching supply use in the early 1960s with the advent of high voltage bipolar transistors, but they weren't economically feasible for low wattage uses until the price of semiconductors became reasonable. Since 1970, advances in all component categories have changed the power supply market to the point where linear power supplies are almost nonexistent above the level provided by three terminal linear regulators. Advances in semiconductors allow single package switching power supplies with multi-watt capability. These designs use the IC, an inductor, and a couple of capacitors to produce a complete voltage regulator in a volume smaller than a single TO-3 switching transistor from the 1960s.

The price per watt of AC line operated power supplies has dropped to the point that it is not cost effective to design and build such a supply in-house unless extremely large quantities are involved. Many companies market lines of standard output voltage supplies. Most of these companies can also supply nonstandard voltages based on standard designs for nominal design fees.

Most of the major linear IC manufacturers (Linear Technology, Maxim, TI, National Semiconductor, Analog Devices, etc.) provide a line of switching regulator circuits suitable for local voltage regulation or voltage conversion. Modern devices from these manufacturers are extremely small and efficient. This is true especially of devices intended for battery-operated equipment where maximum operation between charging is important. Modern devices frequently integrate the control circuit, the switch, and the required rectifiers in the same package.

The passive component manufacturers have been busy improving components as well. The magnetic materials companies (Ferroxcube, Siemens, Micrometals, Magnetics division of Spang & Co., etc.) have extended the useful range of transformers and chokes from the low kHz range (10–50 kHz) in the 60s to well above 1 MHz today. This improvement has allowed much smaller filter capacitors and magnetic cores in modern designs. Capacitor manufacturers have also improved filter capacitors for use in switchers. Ordinary electrolytic capacitors have a very large equivalent series resistance that causes them to dissipate power when a rapidly varying DC voltage is applied. If this equivalent AC current is too high, these electrolytics will heat to the point of explosion. All electrolytic capacitor manufacturers now make lines of capacitors that are designed to limit this equivalent series resistance.

Comparison of Linear and Switching Supplies

A comparison of representative linear and switching power supplies shows why we would want to use a switching supply in most applications.

A linear power supply can only produce a voltage lower than the input voltage. All linear regulators require the input voltage to be at least a minimum amount above the output voltage. This is called the drop-out voltage. The drop-out voltage is the parameter that drives the calculations for efficiency and worst-case power dissipation.

Let's look at the operation of a device that operates at 6.0 V and has a maximum current draw of 2 A. A representative linear regulator will have a drop-out voltage of 2 V. If we choose to use a lead acid battery, the battery will be discharged when the voltage reaches around 1.9 V per cell. Since we require a minimum of 8 V

(6 V for the load plus the 2 V drop-out voltage) for proper operation, we will require a minimum of 5 cells to provide the necessary voltage. This yields a minimum input voltage of 9.9 V when the battery is discharged. The power in the load is 12 W with 2 A supplied, and the regulator must dissipate 7.8 W when the battery is discharged. This yields an efficiency of 60%. When the battery is fully charged, the cell voltage is 2.26 V and the battery supplies 11.3 V. The load power is still 12 W. The regulator must now dissipate 10.6 W, which yields an efficiency of 53%.

The situation is better if we decide to draw less from each cell. We can increase the efficiency and decrease the cost of the battery (at the cost of more frequent recharge cycles) if we stop operation at a cell voltage of 2.0 V. Now we only require 4 cells for operation. The regulator dissipates 4 W at end of charge so the efficiency increases to 75%. At full charge the efficiency has only improved to 67%.

In the first example, 2 of the 5 cells contribute all of their energy to heat. In the second example, 1 of the 4 cells is used entirely for heat. You can see that linear regulation is a very expensive way to provide a constant voltage in a battery-operated system.

A simple switching power supply can be built for the application described above with FET switches that have an on resistance on the order of 0.008 Ohm. The commutating diode can be a Schottky diode with an on voltage of only 0.5 V. As a first approximation, the power dissipated in the switch is a maximum of 0.032 W, and the power dissipated by the diode is 1.0 W. The efficiency at full charge is 92 % and the efficiency at discharge is close to 99%. What is even better is that these relative efficiencies will hold for a 4-cell battery, a 6-cell battery, or a 12-cell battery.

There is another advantage of switching power supplies over a linear supply. With the linear supply, we were restricted to a battery of 4 cells or more for proper operation. A switching power supply can be built to provide the necessary power from 1 to 3 cells that will still have better efficiency than the linear supplies.

The situation is similar for line operated power supplies. A line operated linear supply requires a transformer. A linear supply that delivers 1000 W of power requires a transformer weighing approximately 100 pounds (and heavier if both 50 Hz and 60 Hz operation is required), requires massive heat sinks for the semiconductors and blowers for the heat sinks, and occupies more than a cubic foot of volume. If 110 V or 220 V operation is required, a linear supply will need manual or complicated electronic switching to handle both line voltages. By contrast, a switching supply can be designed that handles 110 or 220 and 50 Hz or 60 Hz without selection circuitry, weighs less than 50 pounds, and occupies one-quarter the volume of the linear supply. The switching power supply also costs a fraction of the linear supply.

Switching supplies are not always the best solution. High frequency noise is an inherent part of the output of a switching power supply. Linear supplies can be 100 to 1000 times quieter than a switching supply. A linear supply is usually a requirement for very noise sensitive analog circuits. Where maximum efficiency is required, modern systems will frequently pre-regulate a voltage with a switching supply to a value just above the drop-out voltage and use a linear supply to provide the low noise power to the analog circuits. Another disadvantage of switching supplies is that there is typically a longer recovery time from a large step change in load current or a step change in input voltage when compared with linear supplies.

Linear supplies are usually a better solution for very low power applications. In the example above, we approximated the loss in the switch as the I^2R power. A better analysis will include losses in the switch during the turn on and turn off times as well as the power needed to drive the switch. Additionally, there are special purpose linear regulators that have very low drop-out voltages for use in low power applications. Both of these factors can tip the balance toward linear regulators in some low power applications.

CHAPTER 1

Basic Switching Circuits

- Energy Storage Basics
- Buck Converter
- Boost Converter
- Inverting Boost Converter
- Buck-Boost Converter
- Transformer Isolated Converters
- Synchronous Rectification
- Charge Pumps

Basic Switching Circuits

In this chapter, we will look at the time domain description of ideal inductors and capacitors and review ideal versions of each type of switching supply. In later chapters, we will look at the magnetic, electrical, and parasitic properties of inductors and capacitors and their effect on the design of individual components.

Energy Storage Basics

Equation (1-1) contains the definition of inductance. An inductor has an inductance of one henry if a change of current of one ampere/second produces one volt across the inductor.

$$V = L \, di/dt \tag{1-1}$$

This is *Lenz's law*. The first consequence of Eq. (1-1) is that the current through an inductor cannot change instantaneously. To do so would generate an infinite voltage across the inductor. In the real world, things such as an arc across switch contacts will limit the voltage to very high, but not infinite, values. The other consequence of Eq. (1-1) is that the voltage across an inductor changes instantaneously from positive to negative when we switch from storing energy in the inductor (*di/dt* is positive) to removing energy from it (*di/dt* is negative). Equation (1-2) is the converse of Eq. (1-1) and is used to determine the current in the inductor when the voltage is known.

$$I = 1/L \int V \, dt + I_{initial} \tag{1-2}$$

Equation (1-3) contains the definition of a capacitor. It states that a capacitor is one farad if storing one coulomb of charge creates one volt.

$$Q = CV \tag{1-3}$$

Equations (1-4) and (1-5) describe a capacitor in terms of voltage and current (where charge is the integral of current and current is dq/dt).

$$V = 1/C \int i\, dt + V_{initial} \qquad (1\text{-}4)$$

$$I = C\, dv/dt \qquad (1\text{-}5)$$

The current waveform of the filter capacitor of a switching power supply is typically a sawtooth waveform. The goal of the capacitor is to limit the change in voltage (ripple voltage). There are two variables in Eq. (1-4) that can control the change in output voltage. We can either make the capacitance large or make dt small to control the voltage ripple. One of the major advantages of switching power supplies is that we can make dt very small (a high switching frequency) which allows the value of C to also be very small.

Buck Converter

Figure (1-1) shows an ideal buck converter regulator made of an ideal voltage source, an ideal voltage controlled switch, an ideal diode, an ideal inductor, an ideal capacitor, and a load resistor. It is called a buck converter because the voltage across the inductor "bucks" or opposes the supply voltage. The output voltage of a buck converter is always less than the input voltage. This ideal regulator is designed to use a 20 V source and provide 5 V to the 10 ohm load. The switch is opened and closed once every 10 µs. The switch produces a pulse width modulated waveform to the passive components. When the regulator is at steady state, the output voltage is:

$$V_{out} = V_{in} * \text{Duty Cycle} \qquad (1\text{-}6)$$

This equation is independent of the value of the inductor, the load current, and the output capacitor as long as the inductor current flows continuously. This equation assumes that the inductor voltage has a rectangular shape.

The diode acts as a voltage controlled switch. It provides a path for the inductor current once the switch is opened. No current flows through the diode while the inductor is charging because it is reverse biased. When the control switch opens, the inductor current flows through the diode.

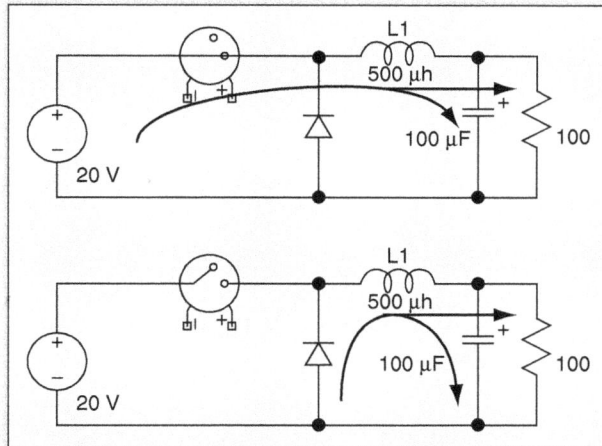

Figure 1-1: Idealized buck converter regulator

We design switching supplies with the simplifying assumption that the applied voltage to the inductor during charging is a perfect rectangular wave. Our example power supply has voltage output ripple of 20 mV. The perfect rectangle is a good approximation since the change in inductor voltage during charging is 0.02/15 or 0.13% and the variation on discharge is 0.02/5 or 0.4%. The constant voltage of the rectangular pulse causes *di/dt* in Eq. (1-1) to be a constant.

Figure 1-2 shows a plot of the output voltage (lower trace) and inductor current (upper trace) after the system is at steady-state providing 5 V and 500 mA to the load resistor.

Note that the change in output current is relatively small compared to the DC value of current in the inductor. In this case, the ripple current is 75 mA P-P. Another important point is that the ripple current is independent of load current when the system is steady-state. This is a consequence of the current through the inductor being controlled by the voltage across the inductor. The slope and duration of charging is controlled entirely by the difference ($V_{in} - V_{out}$). The average inductor current is equal to the output current.

It is also possible for the buck converter to work in discontinuous mode, which means the inductor current goes to zero during part of the switching period.

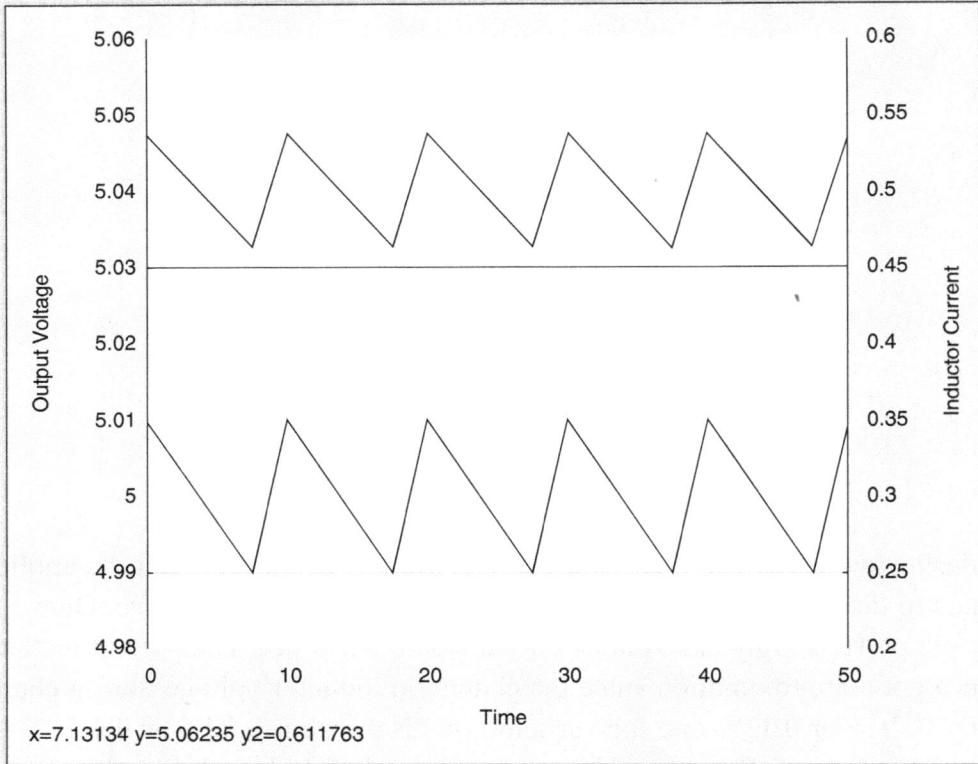

Figure 1-2: Output voltage and inductor current in a buck regulator

Equation (1-6) does not hold for discontinuous operation. The output ripple voltage is higher for a buck converter in discontinuous mode because the capacitor must supply the load current during the time that the inductor current is zero. Usually, a buck converter only runs in discontinuous mode when the load current becomes very small compared to the design current.

Boost Converter

Figure 1-3 shows an ideal boost converter regulator made of an ideal voltage source, an ideal switch, an ideal diode, an ideal inductor, a capacitor, and a load resistor. It is called a boost converter because the voltage across the inductor adds to the input supply voltage to boost the voltage above the input value. The output of a boost converter is always greater than the input voltage. This ideal regulator is designed to use a 5 V source and provide 20 V to the 1000 ohm load. The diode provides a path for the current once the switch is opened. The diode is off while the switch is closed. The switch is opened and closed once every 10 μs.

The switch and voltage source provide current to charge the inductor with energy while the switch is closed. While the inductor is charging, the current in the load is supplied by the capacitor because the diode is reverse biased. When the switch opens, the current in the inductor continues to flow, but now the inductor current forward biases the diode and flows through the load circuit. The voltage across the inductor reverses and adds to the voltage of the input supply. When the regulator is at steady-state, the output voltage is:

$$V_{out} = V_{in}/(1 - \text{Duty Cycle}) \tag{1-7}$$

This equation is independent of the value of the inductor, the load current, and the output capacitor for continuous mode operation.

Boost converters require much more capacitance than a buck converter because the capacitor supplies all of the load current while the switch is closed.

Figure 1-4 shows a plot of the output voltage (lower trace) and inductor current (upper trace) after the system is at steady-state providing 20 V and 20 mA to

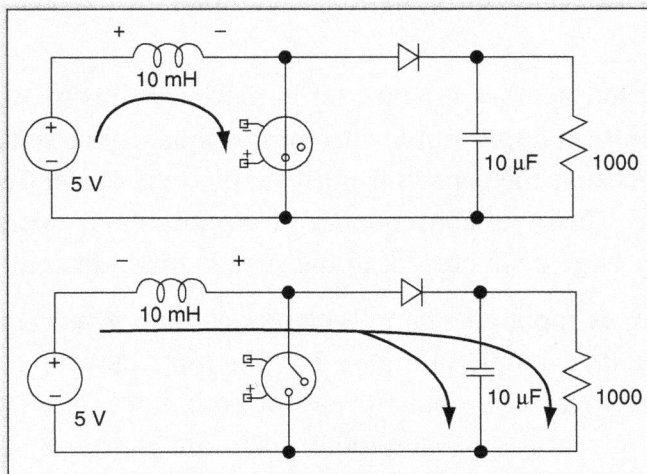

Figure 1-3: Idealized boost converter regulator

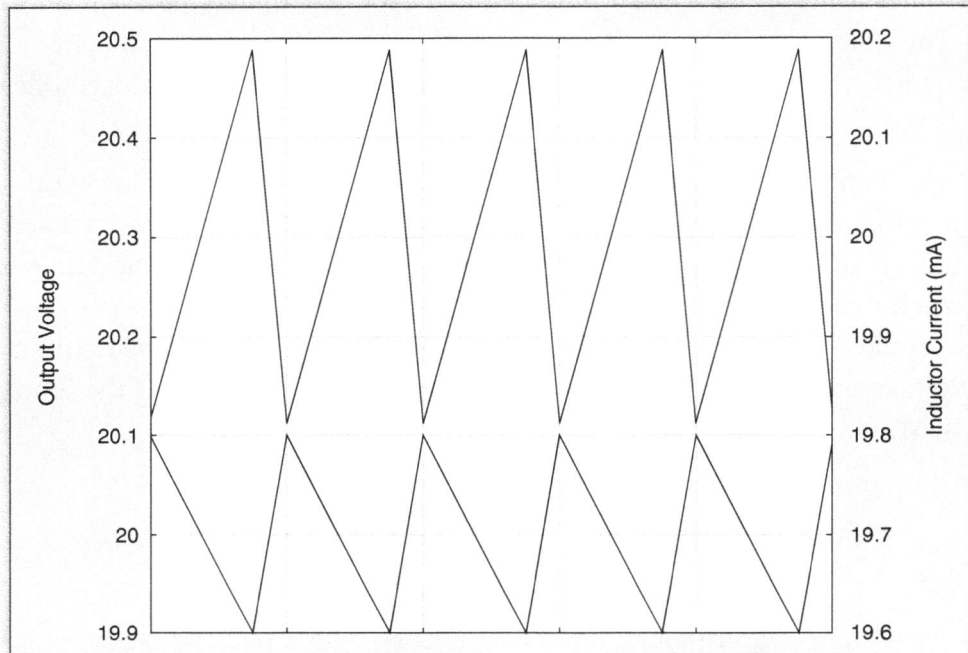

Figure 1-4: Output voltage and inductor current in a boost regulator

the load resistor. Just as in the buck converter, the ripple current in the inductor is independent of the output current for continuous mode operation. Typically, the peak inductor current is only slightly larger than the average inductor current.

It is also possible to run a boost converter in discontinuous mode. Discontinuous mode results in larger ripple current for boost converters, just as in the buck converter, because the capacitor must supply load current while the inductor current is zero. The other consequence of discontinuous operation of boost converters is very large peak current in the switch and inductor.

You can calculate the input current in both modes for a given output current. In our continuous mode example in Figure 1-3, the input current averages 80 mA. Equation (1-8) gives average input current for both modes. Equation (1-9) gives peak input current for discontinuous operation.

$$I_{in\text{-}avg} = I_{out\text{-}avg} \left(1/(1 - \text{Duty Cycle})\right) \tag{1-8}$$

$$I_{in\text{-}peak} = 2 * I_{out\text{-}avg} ((1 - (V_{out}/V_{in}))/\text{Duty Cycle} \qquad (1\text{-}9)$$

If our example circuit had a duty cycle of 0.25 (discontinuous mode) instead of 0.75 (continuous mode), the peak inductor and switch current would be 480 mA instead of 81.75 mA.

Inverting Boost Converter

Figure 1-5 shows the circuit of an ideal inverting boost converter. The switch and voltage source provide current to charge the inductor with energy while the switch is closed. While the inductor is charging, the current in the load is supplied by the capacitor because the diode is reverse biased. When the switch opens, the current in the inductor continues to flow, but now the inductor current forward biases the diode and flows through the load circuit. Since one side of the inductor is tied to the common point, the current flow when the switch opens causes a negative output voltage.

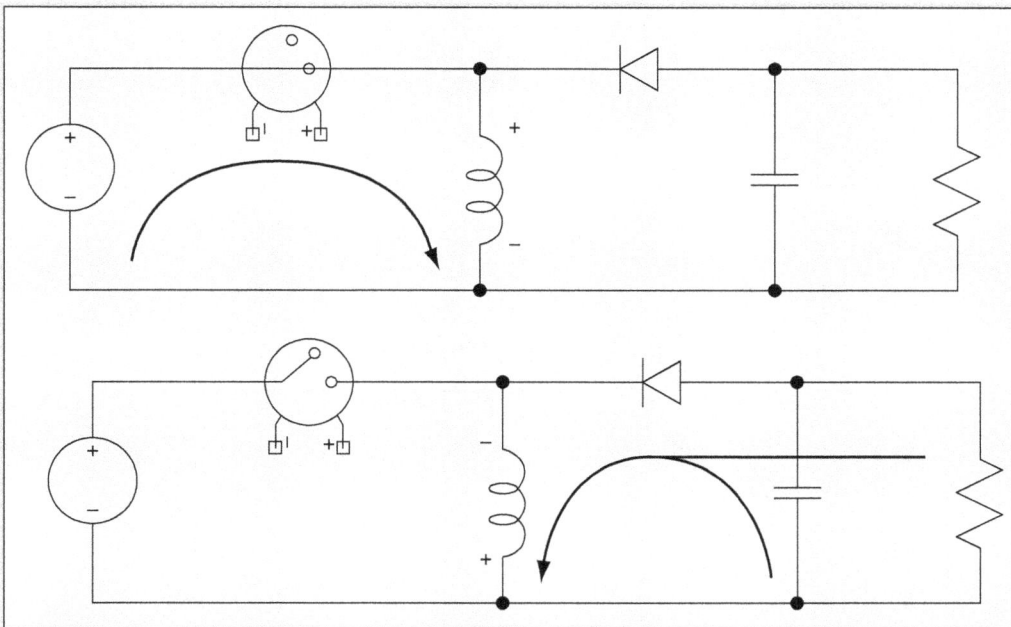

Figure 1-5: Idealized inverting boost converter

When the regulator is at steady-state, the output voltage is determined by Eq. (1-10) for continuous mode operation. Just as in the positive boost converter, the output voltage will be larger in magnitude than (or equal to) the input voltage.

$$V_{out} = - V_{in} * (Duty\ Cycle)/(1 - Duty\ Cycle) \qquad (1\text{-}10)$$

Buck-Boost Converter

If we add an additional switch and an additional diode to the boost converter as in Figure 1-6, we can create a buck-boost converter that will allow us to create a positive voltage that is either above or below the input voltage. Both switches close and open at the same time in this circuit. Again, the inductor is charged while the switches are closed and energy is delivered to the load when the switches open, just as it is in the boost converter. Diode D1 connects one end

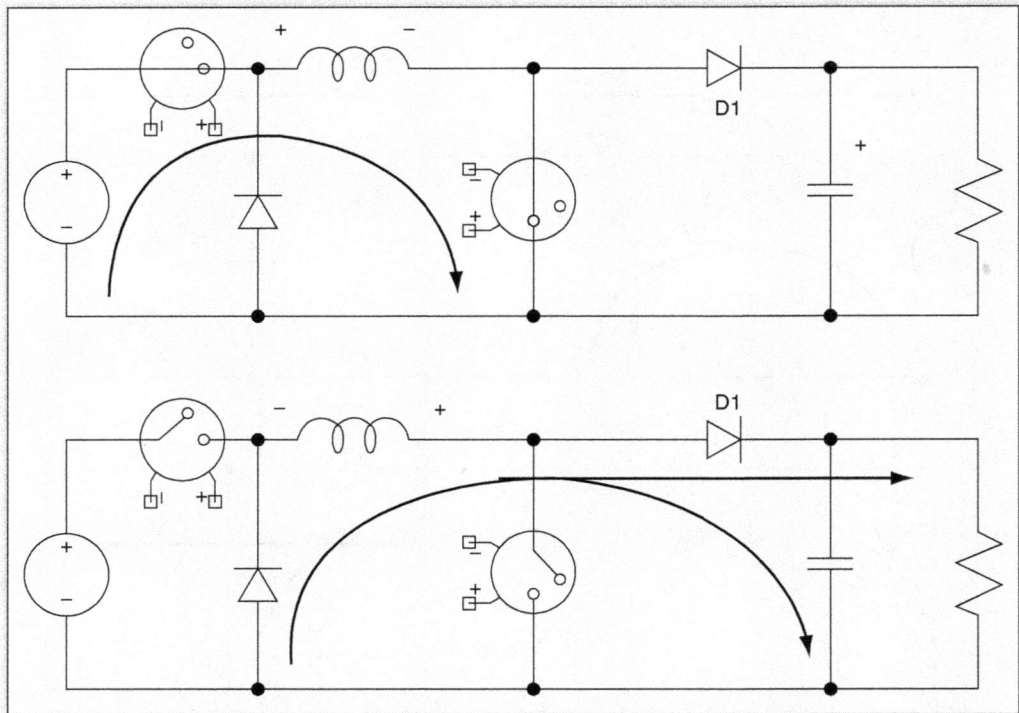

Figure 1-6: Idealized buck-boost converter

of the inductor to the common point so the voltage across the inductor can be either above or below the input voltage.

Transformer Isolated Converters

Power supplies that are intended to run directly from the AC power lines (off-line supplies) require a transformer to isolate the load side from the AC lines. Transformers can also be used in power supplies where isolation is required for other reasons such as medical equipment use. Table 1-1 lists power range and complexity versus appropriate converter type. This table gives a generally accepted range for each converter type. Each type can be used above or below these ranges, but the design problems to create an efficient supply become greater.

An off-line power supply is really a DC power supply that feeds a transformer isolated DC–DC converter. The rest of this section will focus on the DC–DC converter circuits. We will look in detail at the input DC power supply in Chapter 3.

Figure 1-7 shows a single switch flyback converter. It appears that this supply uses a transformer, but, in fact, the magnetic component is an inductor with two windings. This supply uses the primary winding of the inductor to store the magnetic energy in the same way that the boost converter works. Note that the phasing of the windings is opposite to normal transformer use. While the switch is closed, the energy is stored in the core and no current flows in the secondary. When the switch opens, current flows in the secondary [as required by Eq. (1-1)] and delivers energy to the load. The voltage on the output is determined by the turns ratio, just as in an actual transformer. The flyback converter is the only off-line converter that uses an inductor; all others use a transformer. One advantage of the flyback converter is that there is no need for

Table 1-1:

Circuit	Power Range	Relative complexity
Flyback	1W–100W	Low
Forward	1W–200W	Medium
Push-Pull	200W–500W	Medium
Half Bridge	200W–500W	High
Full Bridge	500W–2000W	Very high

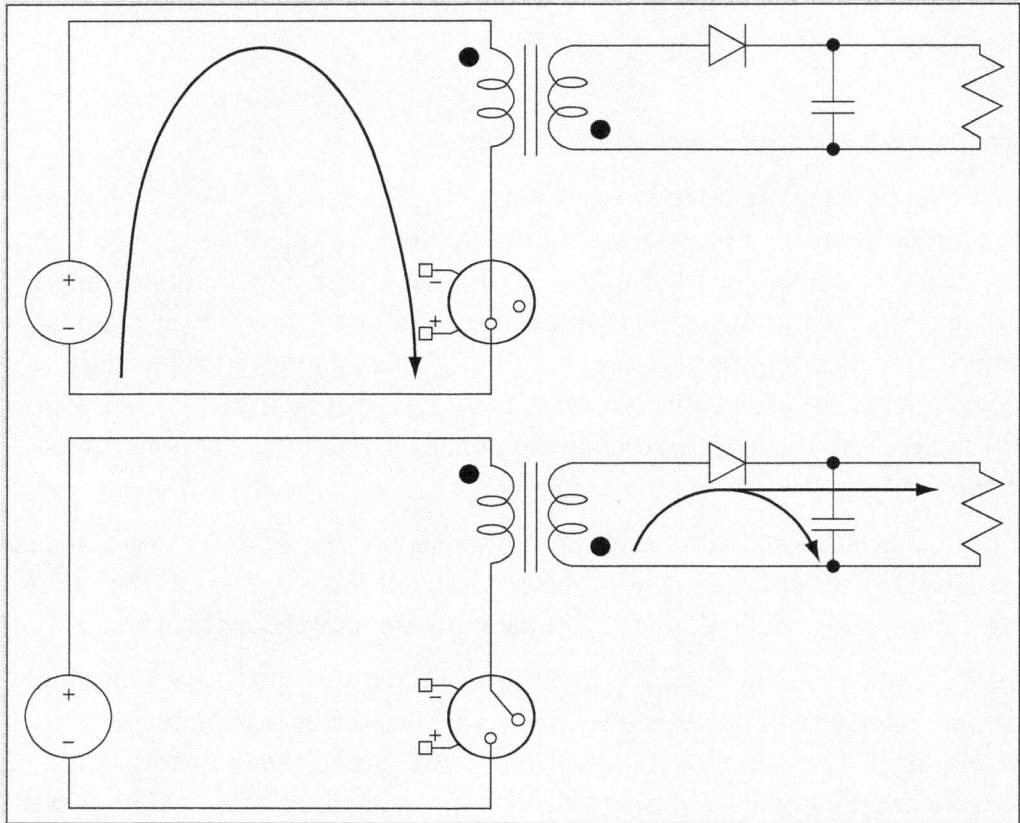

Figure 1-7: Idealized single switch flyback converter

an additional smoothing choke. The energy stored in the inductor is dumped directly into the capacitor and the load. This is also a disadvantage because the current for the load is supplied by the capacitor alone while the inductor is charging. The ripple voltage is larger for the flyback converter unless a larger output capacitor is used.

Figure 1-8 shows a single switch forward converter. During the time the switch is closed, current flows in the primary and in the secondary. The secondary current charges the filter choke just as in a buck converter. When the switch opens, current must continue to flow in the choke, as described in Eq. (1-1). The commutating diode (D2) in the secondary acts just as it does in the buck converter and allows inductor current to continue to flow.

Real transformers also have parasitic inductance that looks like an inductor in series with the primary of the transformer. The primary current that is flowing in the parasitic inductance must continue to flow according to Eq. (1-1) when the switch opens. When the switch opens, current stops flowing in the primary winding and in the secondary winding. The clamp winding (the left one) is phased opposite to the primary and secondary so when the current stops flowing, current begins to flow in the clamp winding as the flux decreases. The current flow in the clamp winding resets the flux in the transformer core to its resting value for the next pulse. The clamp winding acts exactly like the secondary winding of a flyback converter and delivers the energy of the parasitic inductance back to the input supply. There are other mechanisms of resetting the flux in the core, which we will explore in Chapter 5.

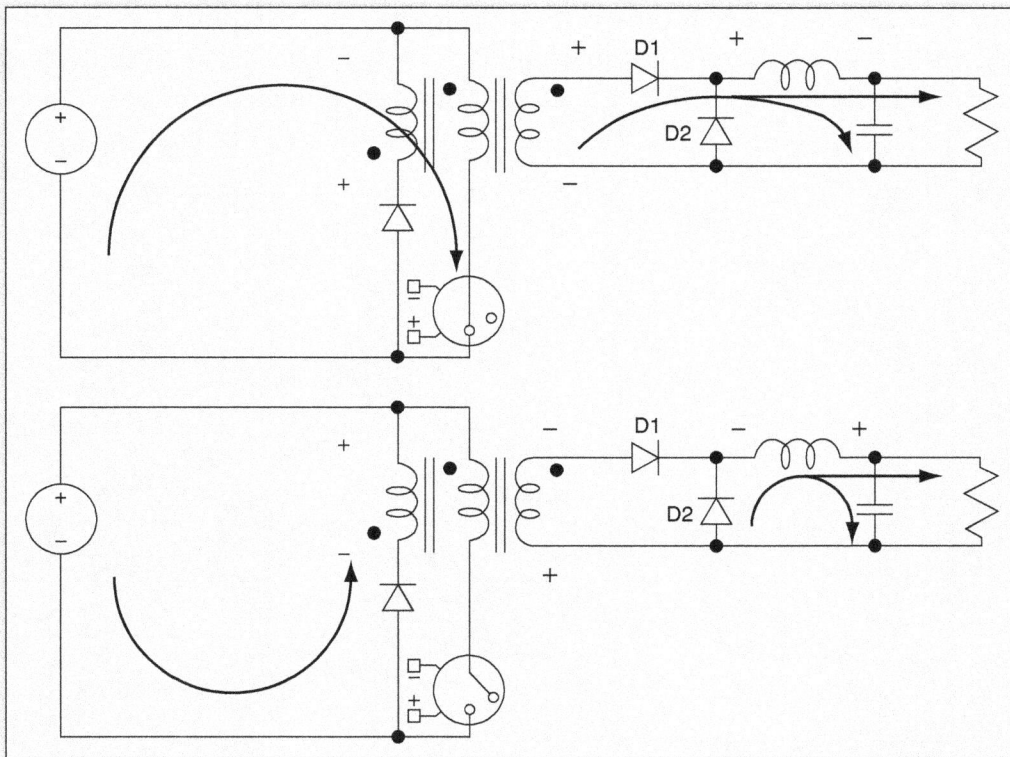

Figure 1-8: Idealized single switch forward converter

Figure 1-9 shows a half bridge converter. This circuit is a high voltage equivalent of a TTL totem-pole output. The switches conduct alternately, which produces a bipolar voltage across the transformer primary. This requires that we have a full wave rectifier for the output. A clamp winding is not necessary since the opposite phase output diode will allow the current to flow in the secondary winding. We can add freewheeling diodes to the primary to control the voltage present on the secondary when the switches open. The capacitors provide a voltage divider that sets one end of the primary winding to one-half the input voltage. These capacitors are almost always part of the input DC power supply, so they perform the dual functions of voltage divider and input charge reservoir.

Figure 1-10 shows a full bridge converter. This design uses four switches to alternate the direction of current through the core.

Figure 1-11 shows a push-pull converter. The switches open and close 180 degrees out of phase, just as in a class B push-pull audio amplifier. Push-pull converters are rarely used in off-line supplies because they require high voltage transistors and it is very difficult to control the flux in the transformer. Modern

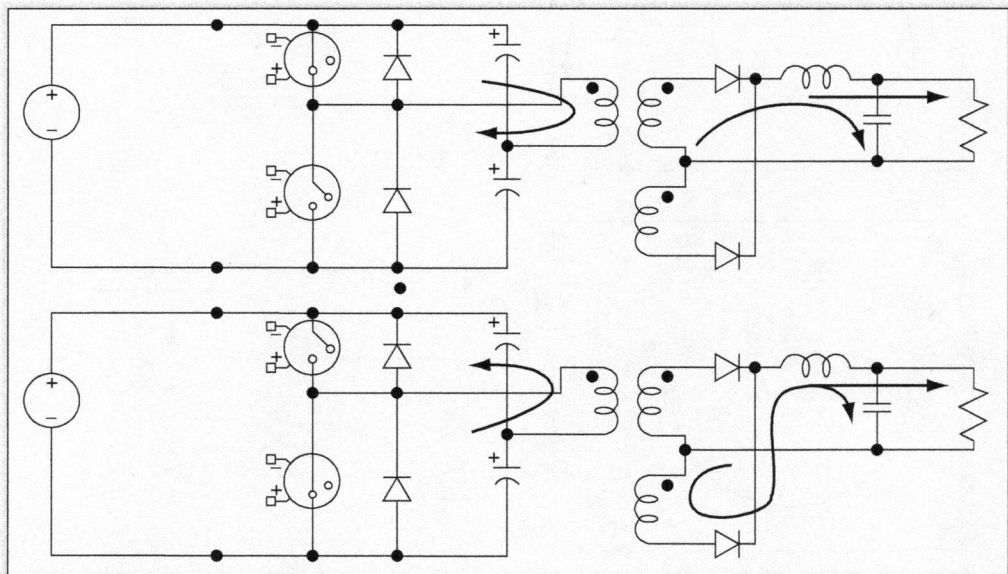

Figure 1-9: Idealized half bridge converter

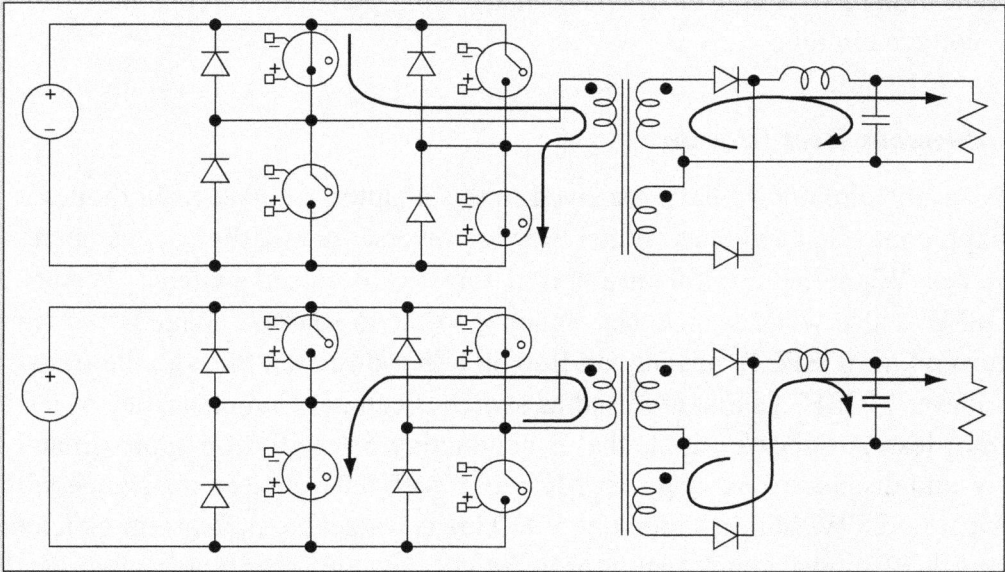

Figure 1-10: Idealized full bridge converter

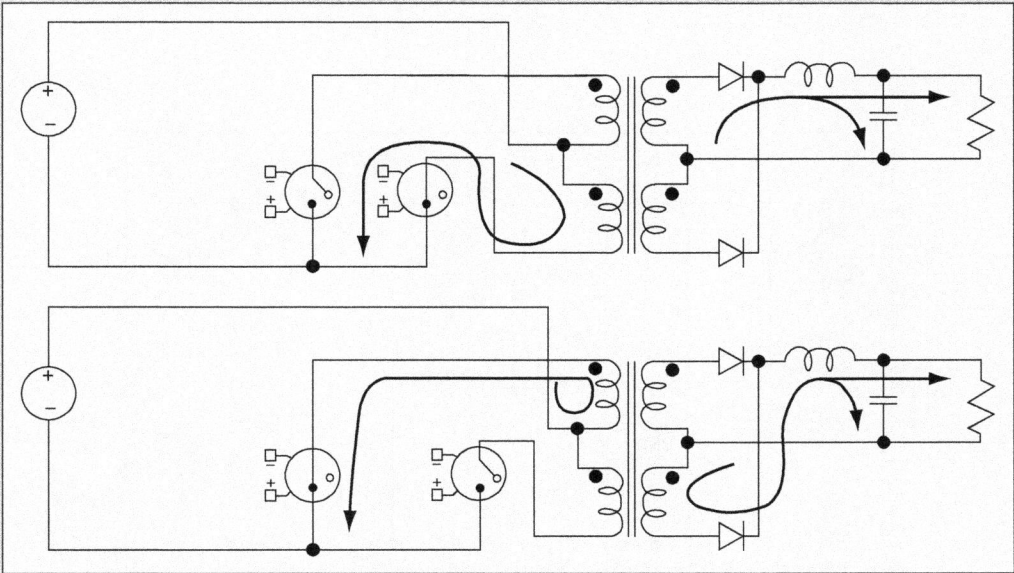

Figure 1-11: Idealized push-pull converter

current mode PWM controllers have made using push-pull circuits practical in low voltage circuits.

Synchronous Rectification

In all of the circuits we have reviewed in this chapter, we have used diodes as voltage-controlled switches. When they are reverse biased, they act as open switches. When they are forward biased, they act as closed switches. Power MOSFETs also work as switches. When the gate to source voltage is sufficient to turn on a MOSFET, current can flow in either direction through the transistor. Power MOSFETs that are used as switches can have on resistance of 0.01 ohm or less. A Schottky diode that is conducting 5 A will drop approximately 0.4 V and dissipate 2 W. A power MOSFET with 0.01 ohm on resistance will dissipate 0.25 W while conducting 5 A. This is a sizeable increase in efficiency. Figure 1-12 shows a buck regulator using synchronous rectification and ideal passive components. This circuit uses an ideal buck converter controller that sequences the MOSFETs and provides the voltage feedback control. When Q1 is on, the circuit turns off Q2. When Q1 is turned off, Q2 is turned on. While this example shows a buck converter, with proper drive circuitry it is possible to replace diodes with MOSFET switches in all designs.

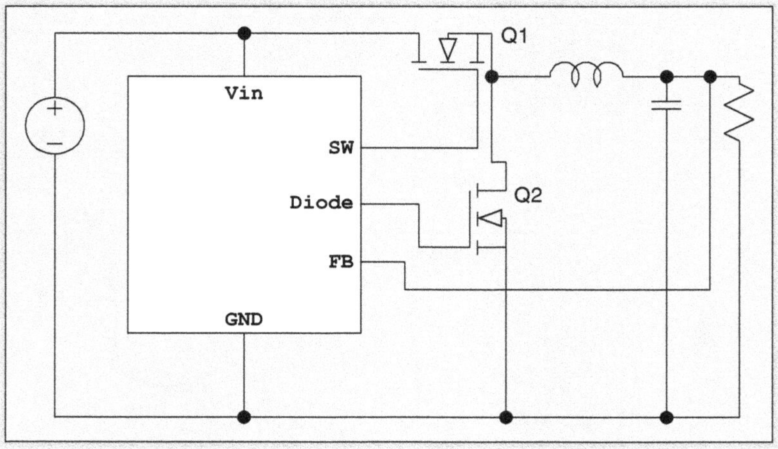

Figure 1-12: Buck converter using power MOSFETs as switches instead of diodes

Charge Pumps

Charge pumps use a capacitor to either increase or invert the input voltage. An ideal voltage doubling charge pump is shown in Figure 1-13. The charge pump capacitor is called a *flying* capacitor (presumably because the switches resemble flapping wings as they change state). During charging, the flying capacitor

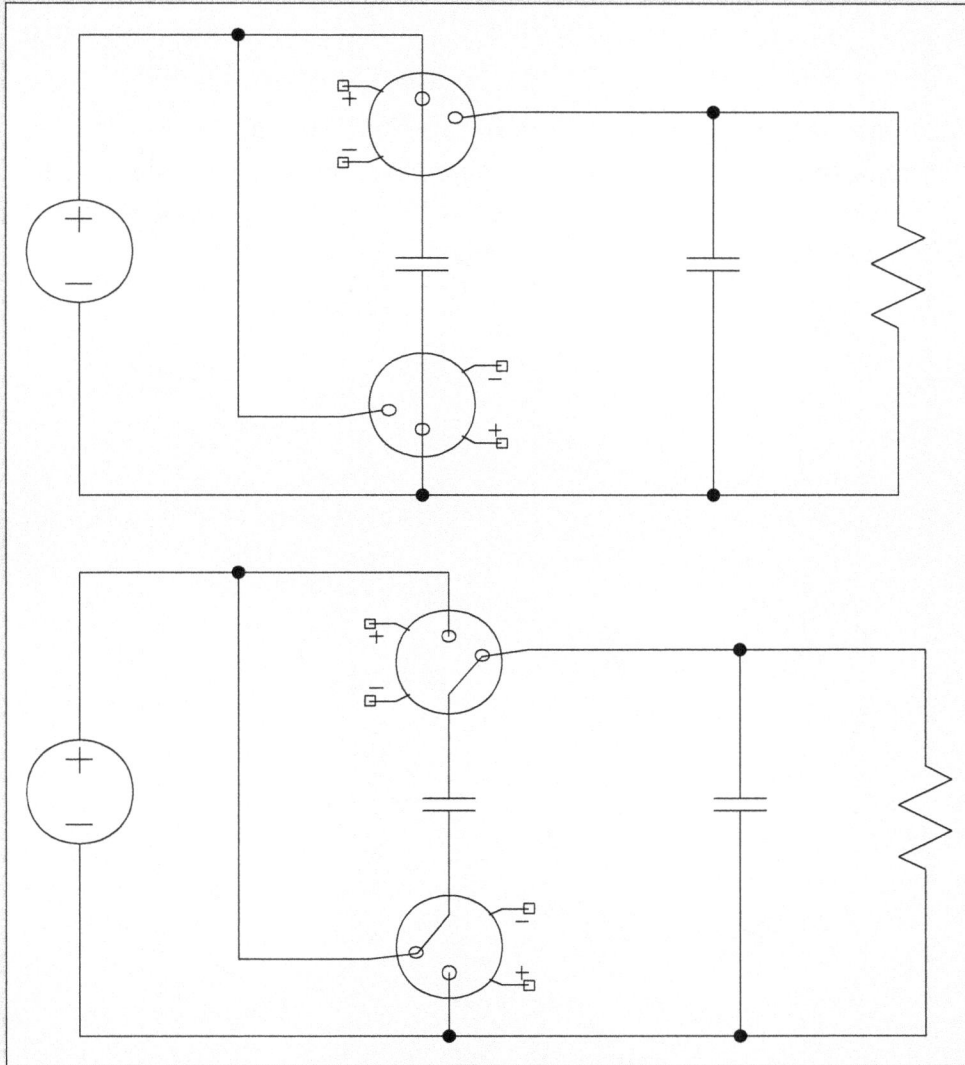

Figure 1-13: Idealized voltage doubling charge pump

17

is charged by the switches. Then the capacitor is connected to the load in series with the input supply to provide a voltage above the input.

Figure 1-14 shows a different arrangement of the switches that allows a charge pump to provide a negative voltage nearly equal in magnitude to the input voltage.

Charge pumps are typically used in applications where a low current is necessary, such as in a bias supply for an IC or a FET amplifier. Charge pumps are not able to supply large amounts of current without using large value capacitors. The practical limit to output current is approximately 250 mA.

A voltage multiplier circuit is also a form of charge pump. Figure 1-15 illustrates a traditional voltage multiplier circuit driven by a totem-pole switch square wave generator. This circuit uses the diodes as switches to steer the current from the generator to the output capacitor.

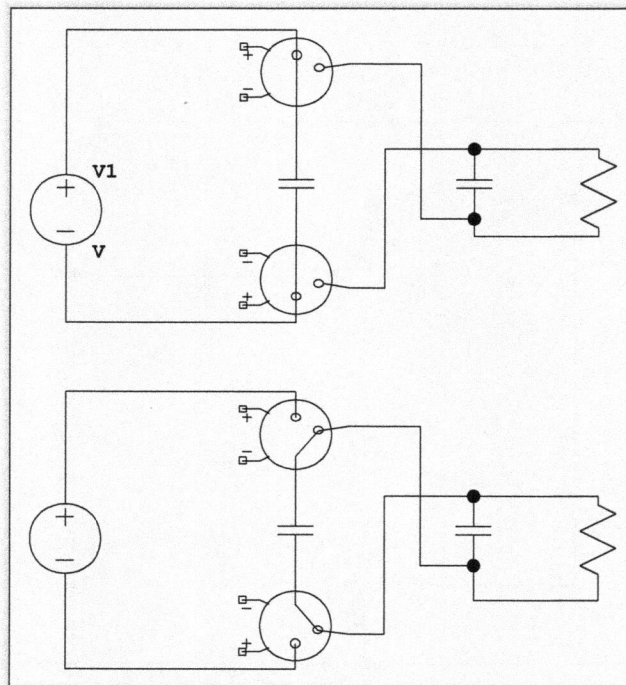

Figure 1-14: Idealized voltage inverting charge pump

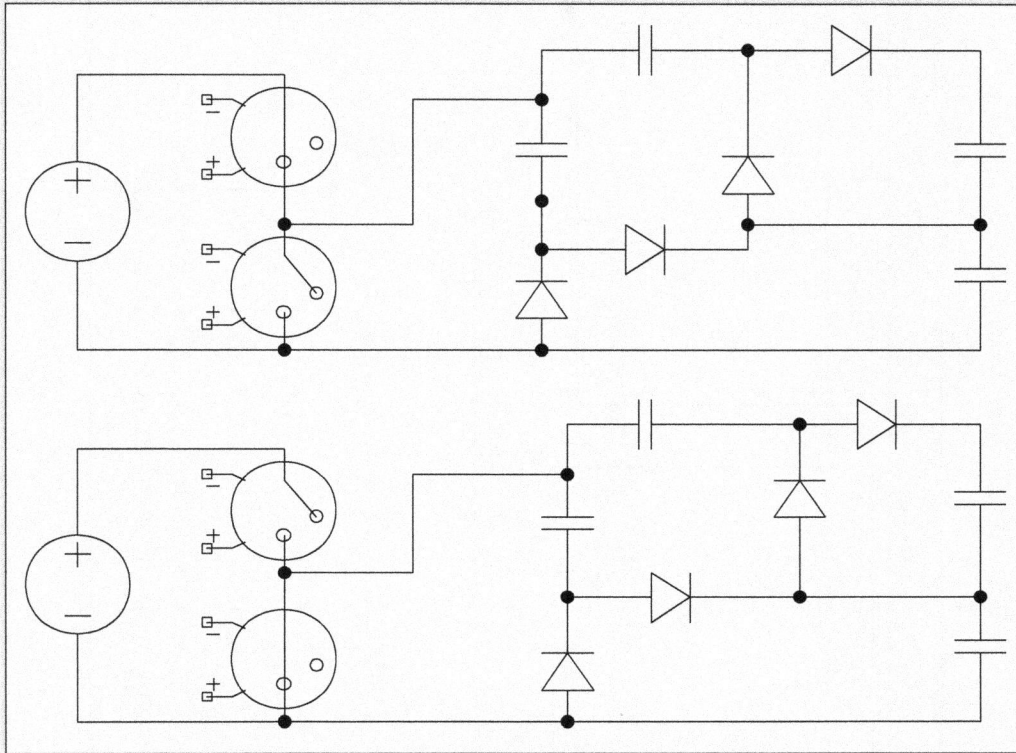

Figure 1-15: Square wave driven voltage multiplier

Figure 1-16 shows a step-down charge pump. This circuit varies the duty cycle to allow the output voltage to be less than the input voltage. The circuits of Figure 1-16 and Figure 1-14 will have an output voltage magnitude that is less than the input voltage. Not all of the energy stored in the flying capacitor can be transferred to the output capacitor. The switching action behaves like an equivalent resistance that depends on the switch frequency and the relative values of the capacitors. We will look at this in detail in Chapter 2.

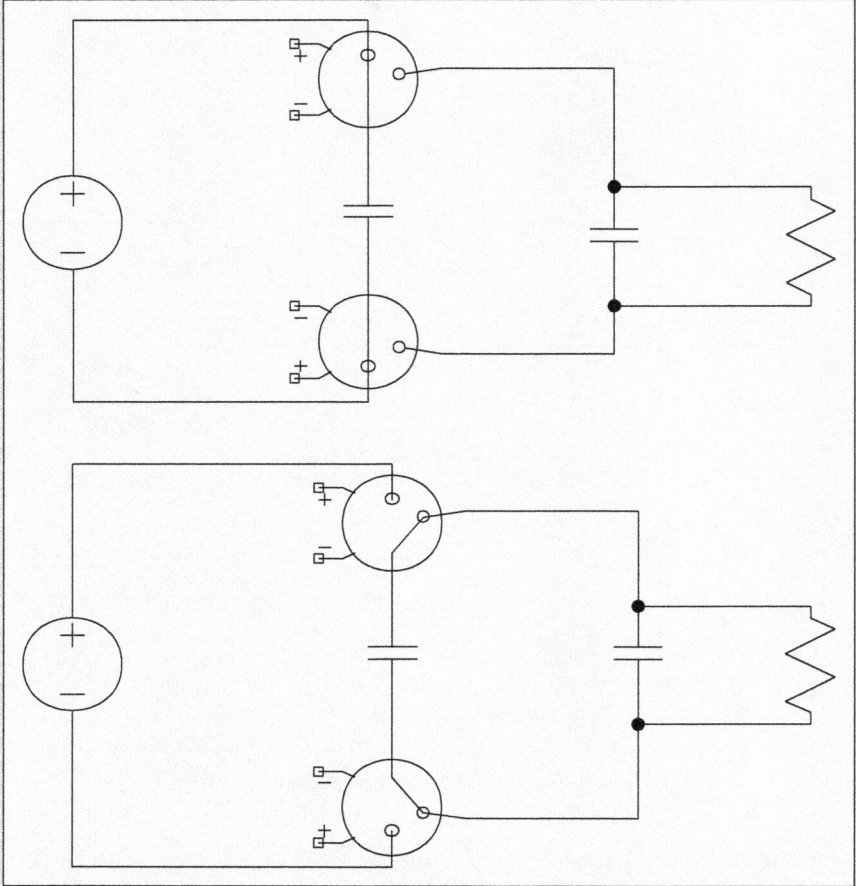

Figure 1-16: Idealized step-down charge pump

Control Circuits

- Basic Control Circuits
- The Error Amplifier
- Error Amplifier Compensation
- Test Sequence
- A Representative Voltage Mode PWM Controller
- Current Mode Control
- A Representative Current Mode PWM Controller
- Charge Pump Circuits
- Multiple Phase PWM Controllers
- Resonant Mode Controllers

Control Circuits

We will explore the various forms of controllers available from semiconductor manufacturers. There is a large variety of controllers available, but each part is usually intended for a narrow application. I will refer to application notes from various manufacturers. These are available on each manufacturer's website or by contacting the manufacturer.

Basic Control Circuits

The simplest form of control circuit is variable frequency/constant on-time or Pulse Frequency Modulation (PFM). In Figure 2-1, the oscillator has a constant on-time (basically a one-shot multivibrator similar to a 555 timer). As soon as the control voltage drops below the reference, the oscillator is triggered to turn on by the comparator. Under light loads, the frequency is low and the duty cycle is low. As the load increases, the frequency increases. The maximum frequency occurs at 50% duty cycle. The wide range of ripple frequency can cause problems for electromagnetic compatibility (EMC) and for ripple control on the output. The Texas Instruments TL-497 is a popular commercial example of this type of circuit.

EMC and ripple control are much more predictable and controllable if a constant frequency is used and the width of the pulse is varied. Pulse width modulation (PWM) uses a constant frequency and varies the on-time of the switch. Figure 2-2 illustrates the basics of a voltage mode PWM controller.

The voltage divider is used with the error amplifier and reference voltage to generate a scaled error signal. The oscillator is similar to a 555 oscillator and generates a constant frequency sawtooth wave. Typically, the timing resistor

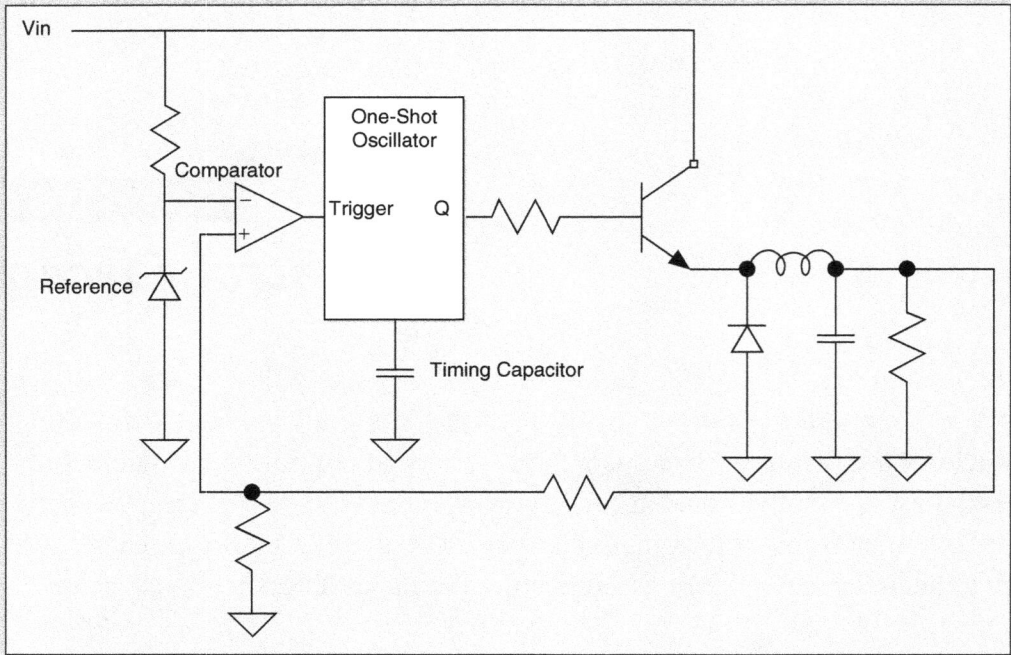

Figure 2-1: Pulse frequency modulation circuit

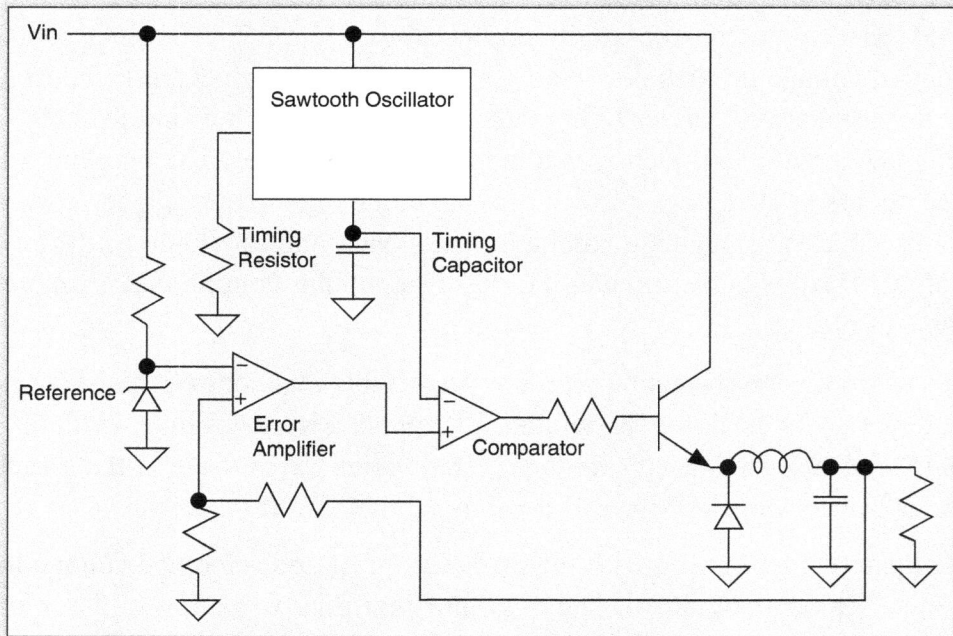

Figure 2-2: Voltage mode pulse width modulation (PWM) controller

sets the charge current for the timing capacitor. Once the voltage on the timing capacitor reaches the trip point, a flip-flop in the oscillator turns on and rapidly discharges the timing capacitor to the lower trip point. The output switch is driven by comparing the error voltage and the oscillator voltage. Figure 2-3 shows how the switch signal is generated.

When the oscillator voltage is less than the error amplifier output voltage, the switch turns on. When the oscillator voltage goes above the error amplifier output voltage, the switch turns back off. If the error voltage is less than the lowest triangle voltage, the duty cycle will be 100%; if the error voltage is greater than the highest voltage of the triangle voltage, the duty cycle will be 0%.

Flyback and boost converters require a minimum amount of off-time so that energy stored in the inductor can be dumped to the output circuit. Some forward converter designs will also require a guaranteed amount of off-time. Modern voltage mode PWM controllers provide a mechanism to ensure a duty cycle less than 100%. This dead time is usually adjustable with an external resistor.

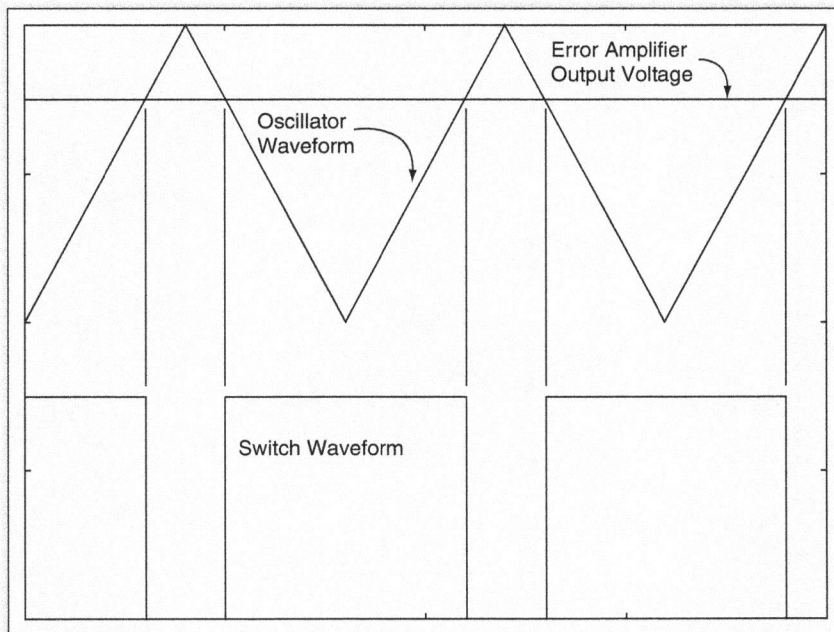

Figure 2-3: Voltage mode switch control generation

Current mode PWM control has inherent advantages over voltage mode control. These include improved transient response and a simpler control loop. Figure 2-4 illustrates the basics of a current mode PWM controller. In this circuit, the oscillator runs at a constant frequency. The pulse from the oscillator sets the flip-flop, which starts current flowing in the transistor switch. The current flow in the switch stops when the current as measured by R_{sense} creates a current sense voltage that equals the trip point set by the error amplifier. The comparator resets the flip-flop, which shuts off the switch. The error amplifier is used to adjust the trip point for the switch current so that the inductor current is the proper amount to maintain the output voltage. As the output voltage approaches the desired value, the error signal reduces the current trip point to maintain a constant average inductor current.

The Error Amplifier

Figure 2-5 shows the typical methods of setting up the error amplifier to control the output for a positive output supply and for a negative output supply. The negative output circuit uses a voltage divider connected to the reference to

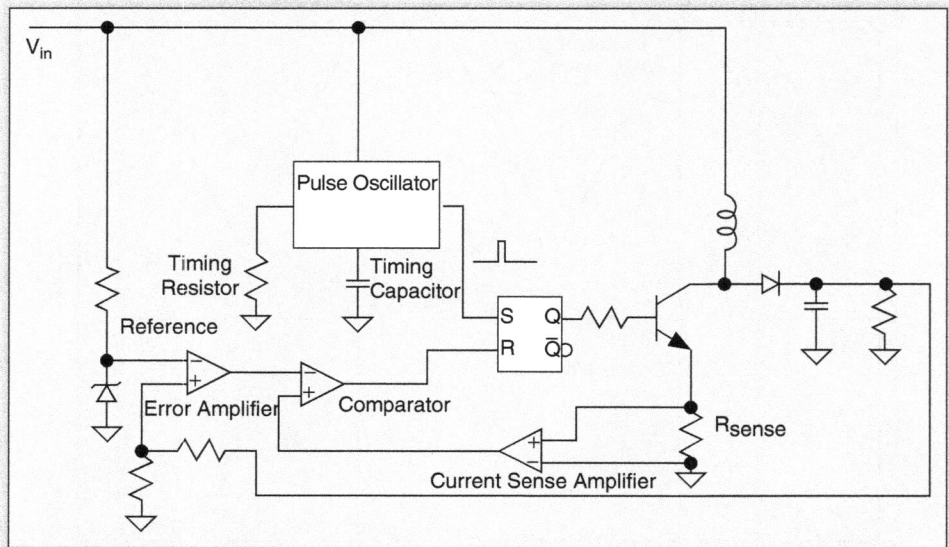

Figure 2-4: Representative current mode PWM controller

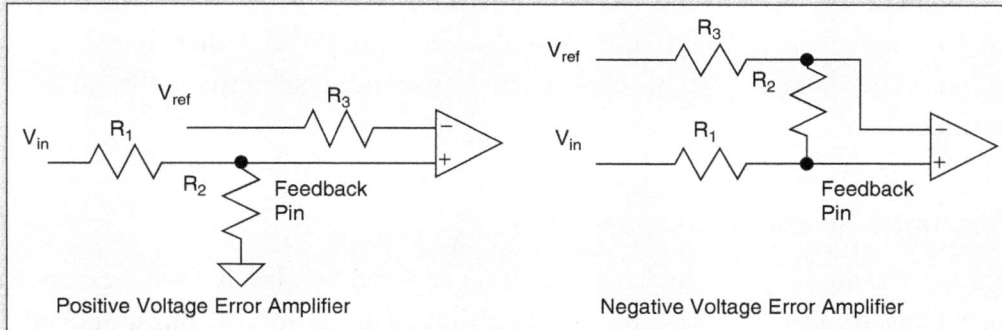

Figure 2-5: Positive and negative voltage error amplifiers

place the input to the amplifier above ground. PWM circuits are intended to operate from a single positive power supply. This means that all of the pins, especially the error amplifier and current sense pins, must not go more than one diode drop below ground.

You will also notice that there is a resistor (R_3) on the pin opposite the feedback pin. All bipolar transistor difference amplifiers (including op-amps and comparators) use the base of a transistor as the input. The input transistors require a small amount of bias current in order for amplification to occur. This bias current flows in R_1 and R_2 in addition to the normal voltage divider current, and it slightly changes the voltage at the feedback pin. The small amount of additional DC voltage due to the bias current will cause a small offset in the output voltage that depends on the closed loop gain of the amplifier and the values of R_1 and R_2. R_3 has a value equal to the parallel equivalent of R_1 and R_2. This ensures that both amplifier input pins are raised above ground by the same amount to balance the effects of the input bias current.

The output of the error amplifier is similar to a resistance-coupled DC circuit. Instead of a resistance, the load for the output transistor is a current source. The effect is that the current is split between the output transistor and the load. This is the equivalent of an open collector digital circuit, except that the transistor is operated in the linear region. Several "open collector" circuits can have their outputs connected together just like a wired-OR open collector digital circuit. The circuit that pulls the output to the lowest voltage is the one that controls the voltage at the input to the PWM comparator. The current source load for the output transistor makes it a transconductance amplifier rather than a voltage amplifier. The voltage gain is equal to the transconductance times the load resistance.

Error Amplifier Compensation

There is a broad class of electronic systems covered by classic feedback control theory. Closed loop op-amp circuits, electromechanical servos, phase locked loops, linear power supplies, and switching power supplies can all be analyzed using control theory. A detailed description of feedback theory is beyond the scope of this book. Thomas Frederiksen gives a very good description of the effects of the transfer function in Chapter 4 of his book *Intuitive IC Op Amps* (National Semiconductor Technology Series, 1984). He describes how multiple poles and zeros can ensure stability or lead to oscillations in a closed loop system. There is also a condensed general description of frequency compensation of amplifier/power amplifier combinations at the end of Linear Technology Application Note 18. Consult a control theory textbook for a complete understanding of compensation.

The error amplifier in PWM controllers is not the equivalent of a 741 or 1458 op-amp. Op-amps have internal compensation that places a low frequency pole somewhere below 100 Hz (usually under 5 Hz). This pole dominates the overall closed loop amplifier performance by rolling off the gain as frequency increases. The error amplifier in PWM controllers usually has no internal compensation. PWM controllers bring the output of the error amplifier out to a pin so that poles and zeros can be added to the closed loop system to provide frequency compensation to the system.

Numerous effects in a switching power supply tend to increase the phase delay around the loop. Two major contributors are the inductor and the filter capacitor, including its equivalent series resistance (ESR). The combination of the inductor and capacitor in the output circuit are the equivalent of a series resonant circuit and will cause two complex poles in the response. The transfer function changes with changes in the load current and power line voltage. The output capacitor and its ESR form a zero, and the load and output capacitor create a pole. Figure 2-6 shows the equivalent circuit of the output capacitor, ESR, and load resistance. You will notice that ESR is a contributor to both the pole and the zero.

The goal of compensation is to ensure that the final power supply will have a quick response to load and input transients, and will not oscillate. Compensation that is heavily damped will guarantee that the output voltage will not oscillate, but the output will likely have a large, long-lasting transient response to rapidly changing input or output. It is also likely to result in significant overshoot during recovery from short circuits. Response that is too rapid will result in oscillations in the control loop.

Figure 2-7 shows a typical compensation network for a buck or forward converter. The resistor and capacitor add a pole to the transfer function. This compensation network needs to be optimized for both gain and frequency. The resistor and capacitor act as a damper to lower the Q of the circuit.

Figure 2-8 shows a typical compensation circuit for a continuous mode boost or flyback converter. All continuous inductor current boost and flyback converters have a zero in the right-half plane. This requires the second pole added to the feedback response. This pole must roll off gain below the frequency of the right-half plane zero. Poles and zeroes in the right-half plane are associated with responses that are steadily increasing in the time domain. The effect of this zero is obvious if you run a simulation of the startup of a boost converter without the second pole. The output voltage will have tremendous overshoot.

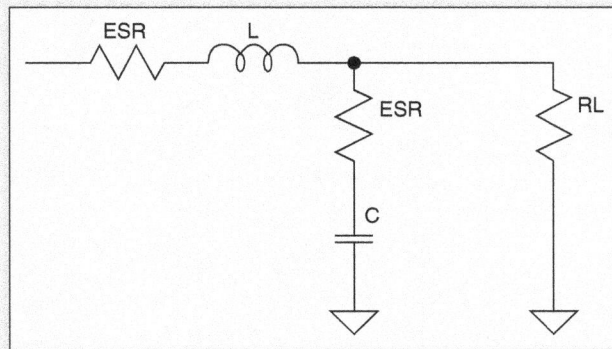

Figure 2-6: Equivalent series resistance in filter inductor and capacitor

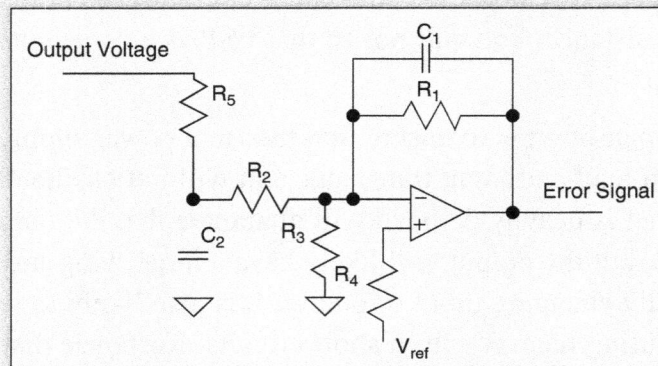

Figure 2-7: Typical compensation circuit for a buck or forward converter

None of the application notes from IC manufacturers gives a rigorous method of evaluating the response of a switching supply using a mathematical approach. Application Note U-95 from Texas Instruments gives some guidance on math for linear power supply compensation that can be used for switching power supply analysis. However, if you understand the math involved, you probably don't need this book.

I prefer the empirical method described in Linear Technology Application Notes 19 and 25 for ensuring that the compensation circuit is optimal for the design. This approach uses time domain analysis rather than frequency domain analysis. The description in these application notes is specific to the LT1070

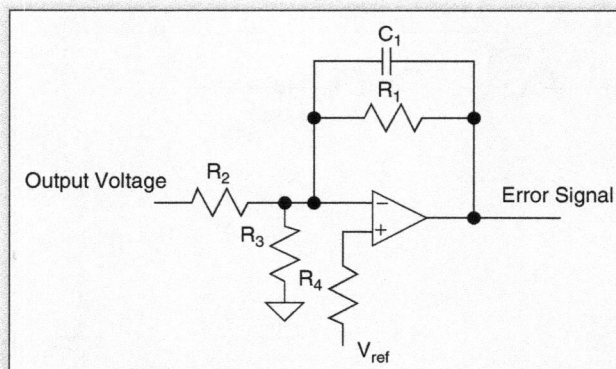

Figure 2-8: Typical compensation circuit for a continuous mode boost or flyback converter

series of current mode controllers, but the technique is applicable to all switching power supplies that have transconductance error amplifiers.

Figure 2-9 shows a test setup based on Linear Technology application notes. There are three pieces of test equipment required. The first is a variable load. This can be an active load that is adjustable or simply a set of high power resistors. The second is an oscilloscope for observing the transient response of the power supply. The last is a function generator that will introduce step changes to the load. We are only interested in the step response, so we place a low-pass filter between the supply output and the input channel of the scope. Thus, we only see the DC value and not any switching frequency energy. We trigger the scope with the output of the function generator.

Test Sequence:

1. Start compensation with a resistance of 1 kΩ and capacitance of 2 μF. This will load the error amplifier for high frequencies and create a dominant pole due to the capacitance and the load of the PWM circuit. There will be a zero in the response due to the resistance but it will have very little effect.

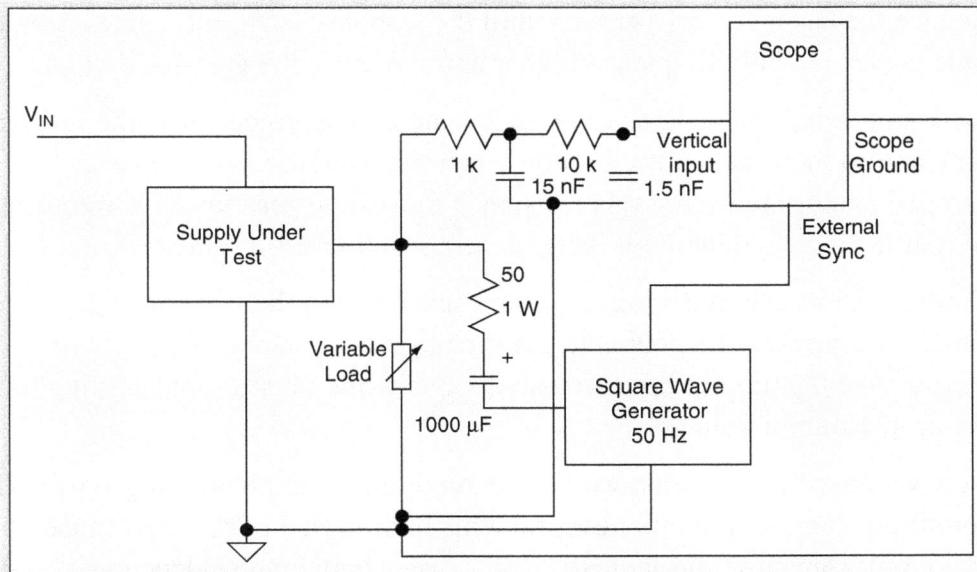

Figure 2-9: Test setup for adjusting compensation in switching power supplies

2. Verify that there are no ground loops by connecting the scope channel 1 probe to the ground connection. If channel 1 shows any response, you must isolate the scope or signal generator by breaking the safety ground connection. In order to maintain safety, you should use an isolation transformer between the test equipment and the power line.

> ***Remember that breaking the electrical safety ground defeats the safety aspect of having a ground. You must use appropriate caution around the test equipment.***

3. Adjust the signal generator for a 5 Vp-p square wave. This gives a 100 mA step input to the control loop. If the positive and negative step responses are not identical with light loading, reduce the signal generator voltage.

4. Verify that the response is a single pole "over-damped" response. If the response is not over-damped, increase the value of the resistor. The resistor should be increased first and then the capacitor value to ensure that we start with an over-damped condition.

5. Reduce the capacitor 2:1 per step until the response is slightly under-damped. This moves the pole frequency higher and increases the gain-bandwidth.

6. Start increasing the resistor value in 2:1 increments to decrease the response time and to increase damping. Stop when the response becomes over-damped again. Increasing this resistance moves the zero lower in frequency so that it starts to flatten out the gain response to mid-frequencies.

7. Continue to iterate, reducing both the capacitor and the resistor to give a fast damped response. The goal is to get the largest resistance and smallest capacitance that do not produce oscillations while giving rapid settling to the proper output voltage.

8. Now we have to verify that we have enough gain and phase margin over all conditions. One of the toughest problems is the value of the zero caused by the output capacitor and its ESR. ESR is very temperature-dependent. If the supply must operate at very low temperatures, the ESR will increase by

orders of magnitude. The margin test will entail testing the response to ensure no oscillations for all combinations of temperature, load, and input voltage. A good rule of thumb is to adjust for slight over-damping at the temperature extremes to ensure stable operation over the entire temperature range.

A Representative Voltage Mode PWM Controller

The 1526A family is representative of a second-generation, full-featured voltage mode PWM controller. This part is suitable for either DC–DC converter service or as an off-line controller at frequencies up to about 100 kHz. This part is especially suited to push-pull, half-bridge, and full-bridge circuits because it has two outputs. Figure 2-10 shows the internal block diagram of the controller.

The internal circuitry requires a stable, regulated voltage for proper operation. The reference regulator is a precision temperature-compensated linear regulator. It is capable of providing 20 mA to external circuits. The reference has a 2 V dropout, so the minimum supply voltage is 7 V. In the 1526A, the band gap reference is trimmed to make the final reference voltage accurate to ±1%.

The under-voltage lockout circuit compares the reference voltage to an internal band gap reference. The circuit pulls the reset pin low, disables the output drivers, and clamps the error amplifier output through the diode so there is no possibility of spurious output pulses until all of the circuitry has sufficient voltage for proper operation. The lockout continues until the reference voltage reaches 4.4 V. The lockout comparator has 200 mV of hysteresis. The circuit will not lock out once the reference reaches 4.4 V until the reference falls below 4.2 V. This prevents noise from causing spurious reset if the reference voltage is rising slowly.

Once the reset pin is released by the under-voltage lockout circuit, the normal soft start sequence begins. The soft start capacitor is connected to the error amplifier output through a clamp transistor that limits how high the error amplifier output voltage can rise during soft start. The clamp on the error voltage limits the maximum pulse width. As a consequence, the increase in inductor current and the rate of rise of output voltage while the system is starting is limited. The clamp is no longer active once the capacitor charges to 5 V. The soft start

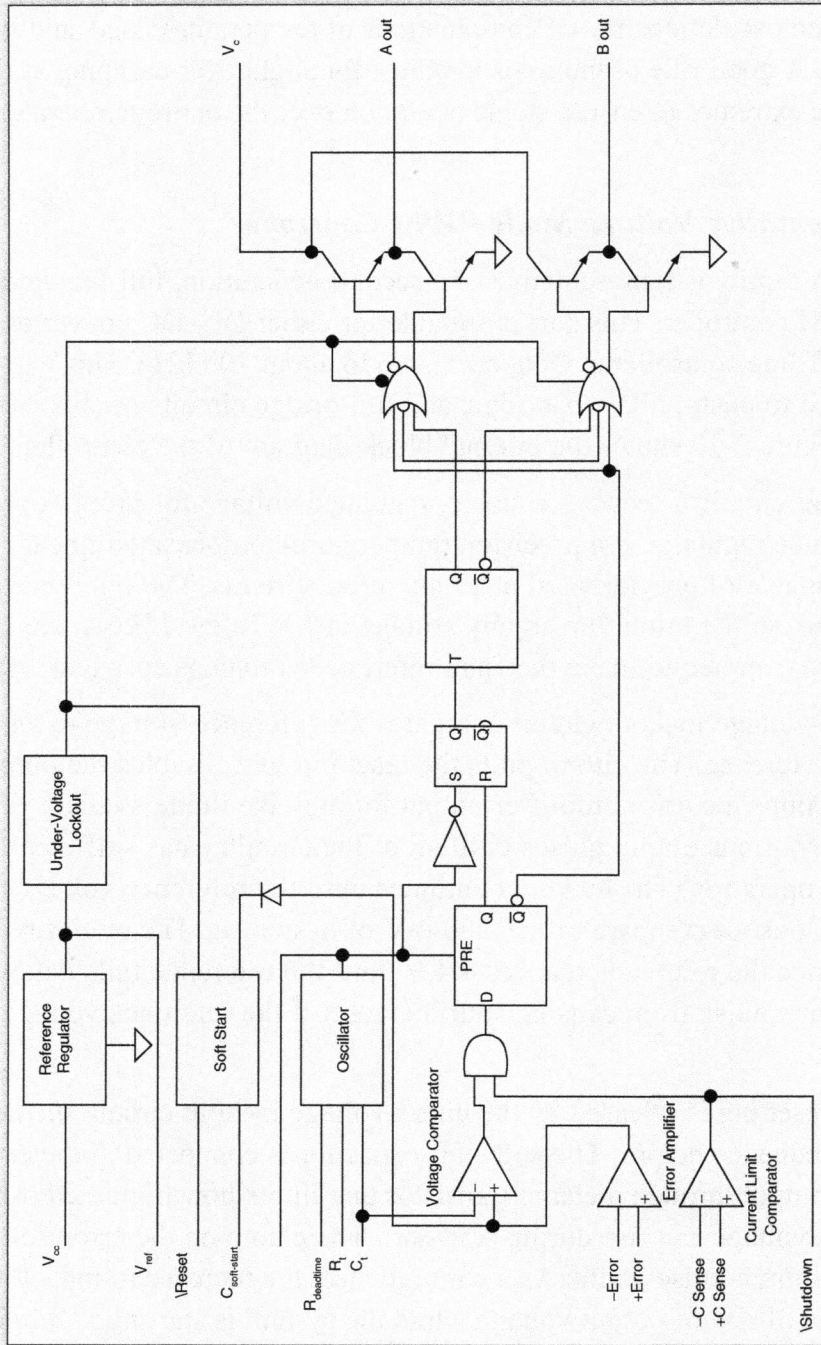

Figure 2-10: Internal block diagram of the 1526A voltage mode PWM controller

capacitor is charged with a constant current of 100 µA (typical), so we can use the capacitor definition and current definition to find the soft start time.

$$Q = C*V \text{ and } I = \Delta Q / \Delta t \qquad (2\text{-}1)$$

If we differentiate both sides of the capacitor equation we get

$$I = C * \Delta V / \Delta t \qquad (2\text{-}2)$$

I is a constant 100 µA and ΔV is 5 V (from reset to fully charged), so we can find the relationship between capacitance and time by rearranging Eq. 2-2.

$$C / \Delta t = 100 \text{ µA} / 5 \text{ V} = 20 \text{ µF/s} \qquad (2\text{-}3)$$

This value is an approximation because the charging current can vary from 50 µA to 150 µA. Also, the normal control loop will begin to dominate the operation of the system long before the capacitor is fully charged.

Soft start is necessary because the current in the inductor is large when the full input supply voltage is across it. It is quite probable that the combination of output capacitor and choke inductance will allow the current to increase so quickly that the output voltage can overshoot the intended voltage by hundreds of millivolts or even several volts. The purpose of the soft start circuit is to protect the diodes and switch transistors from excessive currents during startup and to provide a damped response to the very large transient at startup.

The oscillator in the 1526A provides a dead time control pin in addition to the normal timing resistor and timing capacitor pins. If the R_D pin is grounded, the dead time is controlled by the discharge circuit in the oscillator. Adding a resistor from the R_D pin to ground will increase the dead time. The data sheet lists an increase of 400 ns/ohm when operating at 40 kHz. The data sheet does not give design information for other frequencies, so the value of R_D will need to be determined experimentally. It is obvious from this part of the data sheet that the 1526A was designed when 20 kHz supplies were state of the art. We would want to increase the dead time for push-pull or bridge circuits where we are using slow bipolar transistors as switches. Bipolar switches store charge in the base-collector junction that must be recombined before the transistor will shut off. Increasing dead time ensures that one transistor has completely turned off before the alternate transistor begins to conduct.

The oscillator also has a sync pin that allows the oscillator to be synchronized to an external oscillator or to sync another controller. Some systems contain multiple PWM controller circuits. The sync pin allows all of the controllers to maintain exact frequency and phase so that circuits can be paralleled. The master 1526A is programmed with R_T, R_D, and C_T for the proper frequency. All of the slave 1526 parts share the sawtooth waveform by connecting all C_T pins together. All of the sync pins must also be connected together. All of the slave R_T pins are left open.

The sync pin could also be used to sync the controller to an external logic clock if the system requires. To sync to an external logic signal, you must set the oscillator frequency approximately 10% below the desired frequency. The logic circuit should supply a short pulse (on the order of 500 ns) to the sync pin. This short pulse terminates the charge phase of the oscillator and restarts the cycle.

The sync pin, reset pin, and shutdown pin are all bidirectional, low active logic pins. Figure 2-11 shows how the internal circuits drive the pin as an open-collector output with internal pullup and as an input to the internal circuits. The shutdown pin can be used for fault conditions that require an immediate shutdown of the controller. The shutdown pin doubles as an output indicating

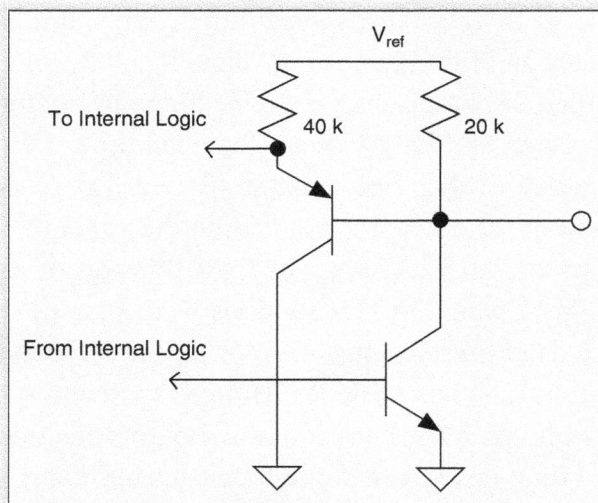

Figure 2-11: Internal circuitry of bidirectional pins in 1526A

36

that the current limit comparator is active. Pulling the shutdown pin low disables the output drivers. The reset pin discharges the soft start capacitor and clamps the error amplifier output. Releasing the reset pin will initiate a soft start cycle. Each of these pins is compatible with TTL or CMOS logic.

The 1526A implements digital current limiting. The current sense comparator provides a logic output that terminates the output pulse. This allows the system to terminate each output pulse if the current limit is exceeded. Do not confuse this operation with current mode PWM control where the error signal controls the current trip point. This part has a fixed threshold for current limit action. The current sense amplifier has an internal 100 mV reference on the inverting pin, so the inverting pin can be grounded to provide for unipolar current sense input. This allows for a very low resistance current sense to minimize current sense power loss.

Other circuits, such as the SG2524, use a difference amplifier that subtracts voltage from the output of the error amplifier and reduces the pulse width output. The internal circuit of the SG2524 is shown in Figure 2-12.

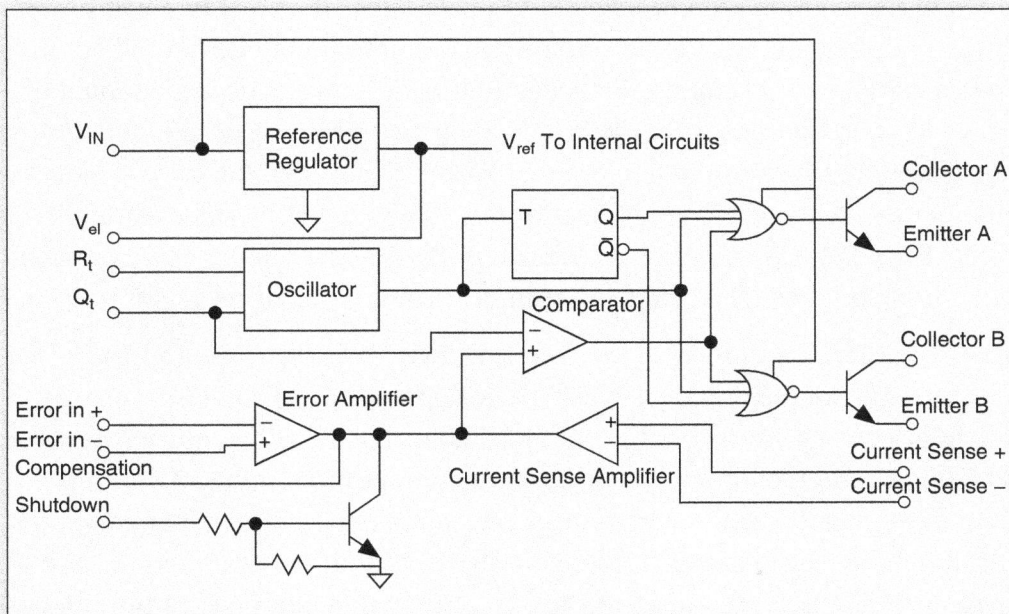

Figure 2-12: Internal circuitry of the SG2524

The 1526A PWM pulse generator uses digital logic to ensure that the comparator does not produce multiple pulses because of noise. The PWM comparator compares the oscillator ramp voltage to the error amplifier voltage and sends a pulse to set the flip-flop when the voltages are equal. The high signal from the PWM latch is sent to the output steering logic. The output pulse is terminated when the oscillator discharge pulse resets the PWM latch.

The output steering logic performs three functions. The first function is implemented by a toggle flip-flop that steers alternate output pulses to alternate output drivers. This allows the 1526A to be used in symmetric drive circuits such as push-pull or bridge circuits. The second function is output blanking. There is a minimum dead time on each output that is controlled by the length of the oscillator reset pulse. The output blanking is ANDed with the pulse command from the PWM latch so that it overrides the PWM latch signal. The third function of the steering logic disables the output drivers for fault conditions such as over-temperature and any time the reset pin is active.

The 1526A has two totem-pole outputs that can be connected to a power supply that is different from the control circuit power supply. This allows the drive to be tailored to the external switches. Each of the outputs is driven at one-half of the oscillator frequency. The pulses from the two outputs do not overlap. When the output is driven low, it saturates the lower transistor. There is a small amount of time where both transistors are on (cross-conduction time) because of the turn-off delay caused by saturation in the lower transistor. Because of the cross-conduction current, this device needs a small resistor in series with the V_C pin to limit the current. The 1526A is an improved version of the 1526 and limits the cross-conduction time to 50 ns. This length of time still requires the current limit resistor.

Figure 2-13 shows a typical drive circuit for FET switches. The 1526A output transistors can source or sink 100 mA. The capacitance of the FET can draw substantial current during charge and discharge. The resistor in series between the FET gate and the output pin protects the output transistors by limiting the peak current. Additionally, the drain to gate capacitance is usually quite large and can couple large inductive voltage transients from the drain circuit into the gate circuit. The Schottky diode ensures that the voltage on the output pin cannot be driven more than 0.3 V negative with respect to the IC ground pin.

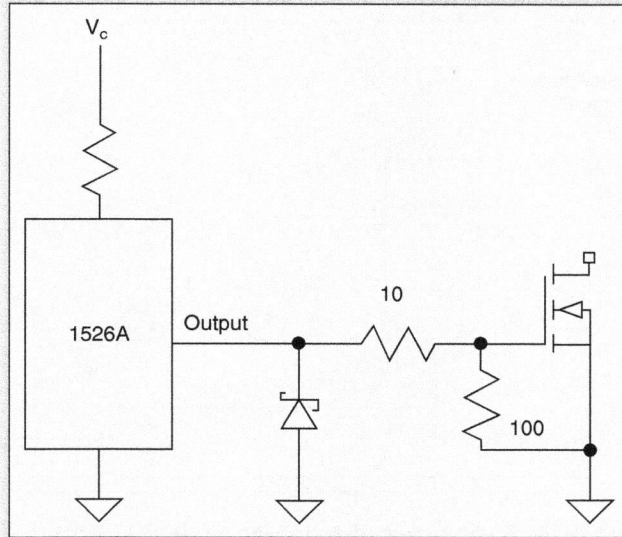

Figure 2-13: Typical drive circuit for FET switches

Current Mode Control

Figure 2-14 shows the basic circuit of a current mode PWM controller in a boost converter. This circuit has two control loops. The outer loop measures the output voltage and provides an error signal to the inner loop. The inner loop compares the error signal and an analog of the inductor current to decide when to turn off the switch. The effect is to change the pulse width. The pulse width is a function of the inductor current rather than a function of the error signal.

The oscillator starts each cycle by setting the output latch to turn on the switch. The error amplifier generates the error signal that is used to compare against the inductor current signal. Once the peak inductor current signal is equal to the error signal, the comparator resets the latch and turns off the switch. If the output voltage decreases, the error signal will increase and allow the peak current to increase with the next pulse.

The operation of the current mode controller has advantages over a voltage mode controller. The first is that the inductor current is a direct function of the error voltage, so for small signal analysis the inductor can be replaced by a

39

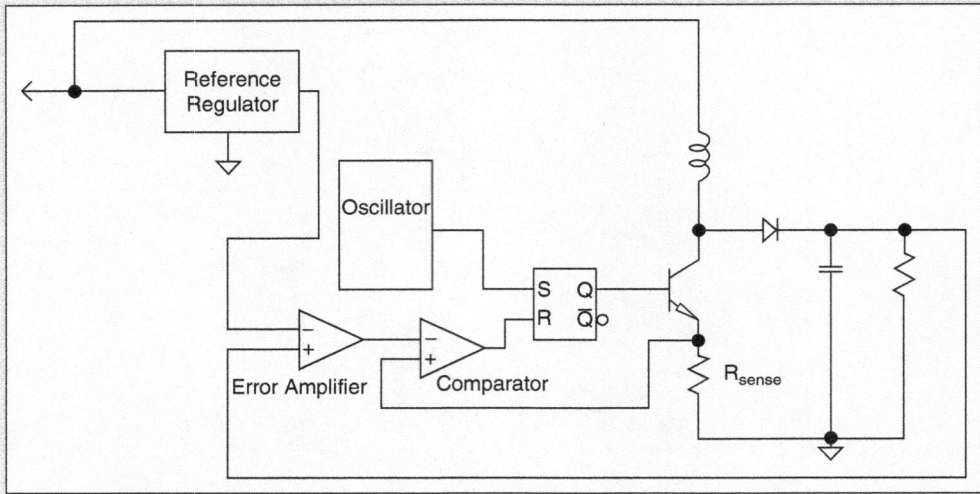

Figure 2-14: Basic circuit of a current mode PWM controller

voltage controlled current source. This removes one order from the transfer function. The control loop is easier to compensate than a voltage mode circuit. Another advantage is that input line voltage changes are removed from the compensation problem. The peak current through the inductor is a function of the voltage across the inductor. If the input voltage drops, it will just take longer for the inductor current to rise to the required value and for the comparator to shut off the switch.

Current mode controllers are not without their problems. Whenever the duty cycle exceeds 50% and inductor current is continuous, current mode controllers have a response called subharmonic oscillation. The inner current loop is unconditionally stable as long as the duty cycle is below 50%. When the duty cycle is larger than 50%, the output will diverge from stable control when the inner loop is perturbed by noise or transients. The average inductor current will stay in control and be set by the error amplifier, but it will vary at subharmonics of the switch frequency. For a 40-kHz switch frequency, the inductor current will have frequency components at 20 kHz, 10 kHz, etc. These subharmonic frequencies can produce audible responses in the inductor and other components. A current mode controller can be stabilized to maintain control by adding slope compensation. Slope compensation is usually accomplished by feeding some of the voltage from the oscillator capacitor into either the current sense amplifier or the error amplifier. Slope compensation changes the current trip from a constant voltage to a sawtooth waveform at the switch frequency. The trip current decreases as the duty cycle increases. There is a minimum compensation slope that will guarantee that the system is unconditionally stable. The inequality below describes this relationship:

$$S_{COMPENSATION} \geq S_{CHARGE} \, (2 \, DC - 1)/(1 - DC) \qquad (2\text{-}4)$$

$S_{COMPENSATION}$ is the slope of the compensation voltage and S_{CHARGE} is the slope of the inductor charging waveform. Fortunately, most modern current mode ICs provide internal slope compensation that can be used "as is" or modified if necessary. For older parts, such as the 1846A, either a manufacturer's application note or the data sheet will give the information necessary to calculate the appropriate amount of slope compensation. TI Application Note U-97 and Linear Technology Application Note 19 give detailed analyses of slope compensation.

A Representative Current Mode PWM Controller

The 1846A is representative of a third-generation controller. Figure 2-15 illustrates the internal circuitry of the 1846A. The oscillator and reference are basically the same circuit as used in the 1526A. The 1846A oscillator can be synchronized to another 1846A or to an external oscillator in the same way as performed on the 1526A. The under-voltage lockout circuit is different in that it uses the input voltage to make the lockout decision rather than the reference voltage. The under-voltage lockout holds the device in reset as long as the input voltage is below 8.0 V. The lockout circuit has 0.75 V of hysteresis to ensure that noise or slowly rising input voltage will not cause an unstable condition at turn on.

The error amplifier is a transconductance amplifier with an "open collector" output similar to that in the 1526A.

The current sense amplifier is a voltage difference amplifier with a gain of three. The diode and voltage source in series with the inverting input of the

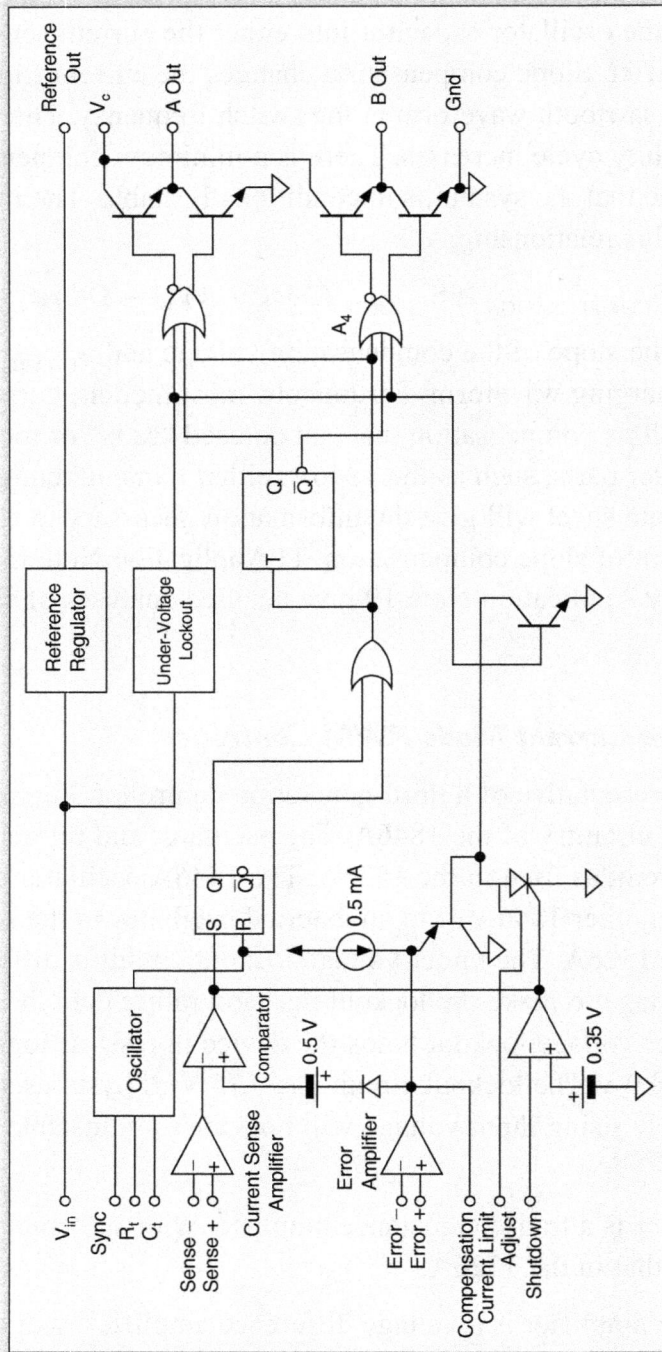

Figure 2-15: Internal circuitry of the 1846A PWM controller

PWM comparator limit the voltage to about 3.5 V (4.6 V error signal max. minus 0.5 V minus one diode drop). This means that a current sense amplifier output above 3.5 V will not shut off the output pulse. This constrains the current sense voltage to be less than 1.1 V because of the gain of three in the current sense amplifier.

The inverting and noninverting inputs have a common mode range of ground to $V_{IN} - 3$ V. This allows the current sense amplifier to be used in boost, buck, forward, and flyback designs. Figure 2-16 shows three different methods of implementing current sense. The resistor and capacitor in Figure 2-16(a) are usually necessary to reduce the size of turn-on transients in the switch. In both bipolar and FET switches, there is coupling between the high voltage side of the switch (collector/drain) and the current sense resistor. The transient that couples to the current sense resistor can cause a false termination of the output pulse. The resistor and capacitor limit the rise time and reduce the transient so proper operation occurs. Buck designs will require that the input voltage is at least 3 V above the output voltage if a current sense resistor is used. In circuits where there is not sufficient common mode range or when total isolation is required (as in bridge circuits), the current limit amplifier can be driven by an isolating current transformer. A current transformer is also advantageous in very-high-current applications because it can reduce the voltage and, therefore, the power consumed by the current sense. The diode in figure 2-16(c) is necessary so that the voltage at the noninverting amplifier input does not go more than one diode drop negative from ground.

The shutdown circuit, the under-voltage lockout circuit, and the current limit circuit clamp the output voltage of the error amplifier. The current limit pin is

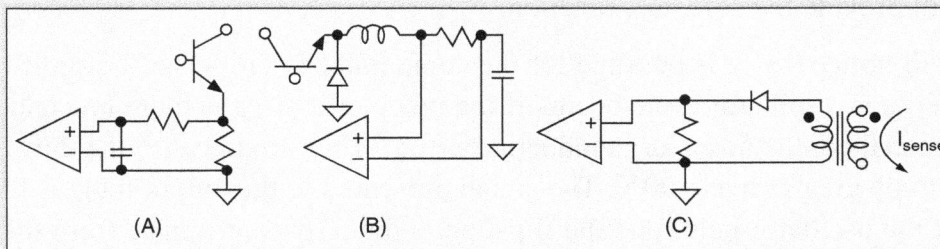

Figure 2-16: Three different methods of implementing current sensing: (a) grounded resistor; (b) floating resistor; and (c) with an isolating current transformer

used to limit the maximum inductor current by clamping the error amplifier output below the 4.6 V maximum of the error amplifier. The error amplifier output is clamped to a voltage equal to one diode drop (the base-emitter voltage) of the current limit set transistor. Figure 2-17 shows a typical connection to the current limit pin. The current limit is not directly set by the voltage at the current limit pin, but, rather, it sets the current sense output voltage that will terminate a pulse. Since the diode drop in series with the inverting comparator input is roughly equal to the base-emitter voltage, the trip point is equal to the current limit voltage minus the 0.5 V offset. The equations below allow you to set the current limit:

$$V_{CURRENT\ LIMIT} = R_1/(R_1 + R_2) * V_{REF} \qquad (2\text{-}5)$$

$$V_{CURRENT\ SENSE} = (V_{CURRENT\ LIMIT} - 0.5)/3 \qquad (2\text{-}6)$$

$$I_{CURRENT\ LIMIT} = V_{CURRENT\ SENSE}/R_{SENSE} \qquad (2\text{-}7)$$

R_2 has a secondary function of supplying the holding current for the shutdown latch. If you desire the shutdown to latch, R_2 must be below 2.5 kΩ to supply at least 1.5 mA of current to hold the latch. When the shutdown signal goes below 350 mV, the shutdown circuit will shut off the PWM latch and hold the IC in reset until a power cycle occurs. Selecting R_2 greater than 5 kΩ will allow the shutdown circuit to reset the PWM latch and discharge any capacitance on the current limit set pin, but when the shutdown signal is removed, a new start sequence will begin.

This IC does not provide soft start circuitry. Soft start is accomplished by adding a capacitor to ground on the current limit pin. The current limit pin sets the peak current trip point, so raising the voltage slowly on the current sense pin will provide the soft start function.

You will notice that it is possible for the comparator to fail to set the flip-flop before a new oscillator cycle occurs if the inductor current is quite low and the error signal commands a large inductor current. This would cause the duty cycle to be greater than 100%. The signal presented to the output logic is the OR of the oscillator pulse and the flip-flop output. The short pulse from the oscillator will guarantee a short dead time on the output equal to the discharge time of the timing capacitor. You can adjust the length of the dead time by

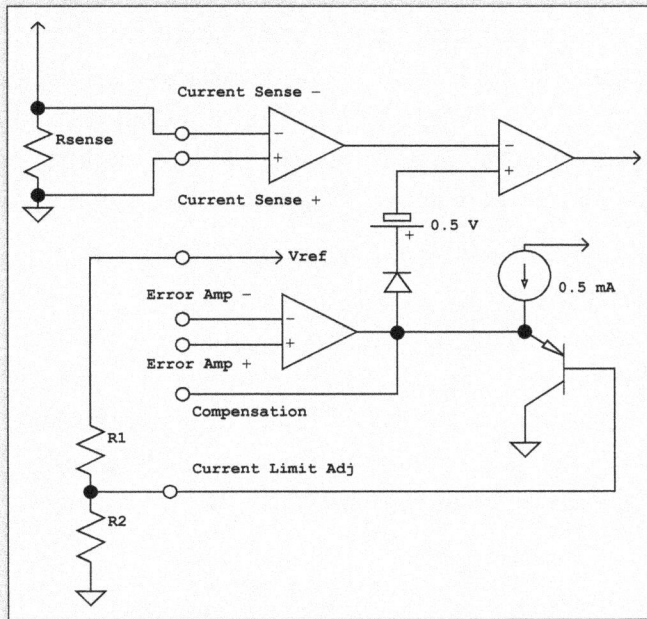

Figure 2-17: 1846 current limit pin implementation

changes to the relative values of the timing resistor and capacitor. The data sheet gives a nomograph for setting the dead time.

The output logic and totem-pole outputs of the 1846A are similar to that of the 1526A. You must follow the same precautions of limiting the current into the collector supply during the crossover in the output transistors. You must similarly limit the output current when driving FET switches by using series resistors.

Charge Pump Circuits

IC manufacturers continue to improve the output capability of charge pump converters. Switch frequency and switch on resistance are the two parameters which affect power dissipation and indirectly affect efficiency and maximum output current. Charge pump circuits have an equivalent series resistance that is given by:

$$R_{EQ} = 1/F_{SWITCH}C_{FLYING} \qquad (2\text{-}8)$$

This equivalent resistance is a property of the switched capacitor circuit and is not an actual physical resistance. You can see that we can improve performance (lowering R_{EQ}) by raising the frequency or increasing the flying capacitor. Performance is only increased until the internal physical resistance of the switches approaches the equivalent resistance of the switching circuit. In general, charge pump ICs can be used in parallel to achieve greater output current.

Figure 2-18 shows the internal circuit of the LTC3200, which is a representative voltage doubling charge pump that provides a regulated output. The circuit contains a 2-MHz fixed-frequency oscillator that drives the switch circuit with a two-phase nonoverlapping clock. The error amplifier compares the voltage at the feedback pin to the internal 1.268-V zener voltage reference. The output of the error amplifier controls the amount of current that can flow into the flying capacitor during phase one of the clock. Phase two of the clock connects the

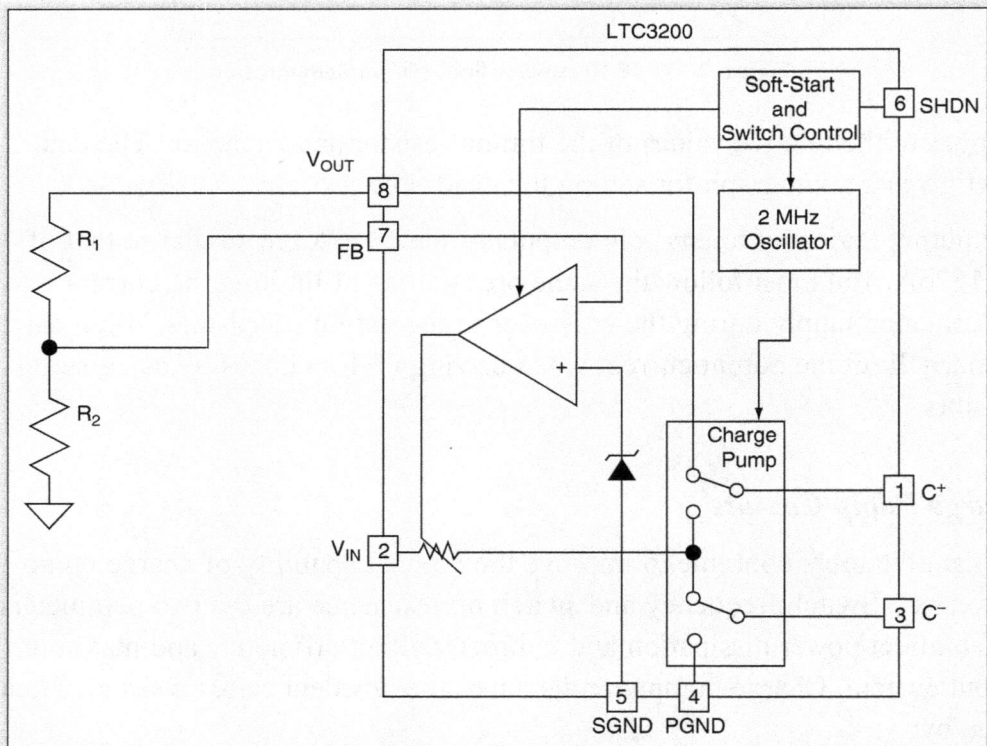

Figure 2-18: Internal circuitry of the LTC3200

flying capacitor in series with the input voltage to provide current to the load and output capacitor.

This charge pump controller has a soft start and switch control circuit to limit the current draw from the input supply. The switch control circuit shuts down the system if the IC gets hotter than 160°C and reenables the circuit around 150°C. This circuit also limits output current to 225 mA in the case of a short circuit.

The LTC3200 will produce a regulated output between 1.268 V and 5.5 V, up to 100 mA. The input voltage range is 2.7 V to 4.5 V. This can accommodate a single lithium cell, three alkaline cells, three NiCad cells, or three NiMH cells. The current control circuit allows the IC to regulate the output to a voltage above or below the input voltage. Efficiency suffers for output below the input, however. The output voltage is set with a voltage divider between the output pin and the feedback pin. The equation for the output voltage is:

$$V_{OUT} = 1.268 \left(1 + (R_1/R_2)\right) \tag{2-9}$$

The resistors can range from several kilohms to 1 MΩ. If the output voltage will be below the input voltage, it is necessary to put a 1 mA load on the output to ensure that the voltage does not creep up at very light loads.

The input capacitor, output capacitor, and flying capacitor are all required to have low ESR. These capacitors need to be greater than 0.5 μF, but values up to only 1 μF will be adequate for good output current and low ripple. Electrolytic and tantalum capacitors will not have low enough ESR to work properly. Ceramic capacitors are the preferred type. Ceramic capacitors have a significant temperature coefficient depending on the type of dielectric. X5R and X7R capacitors have the smallest change in value over temperature. Another consideration is the change in capacitance with applied voltage. Z5U and Y5V capacitors have significant changes in capacitance with applied voltage. The output capacitor ESR must be below 0.3 Ω in order for the error amplifier to remain stable. If the ESR is higher, the amplifier response is no longer a single pole rolloff and can become unstable.

The LTC3200 uses a variable resistance to modulate the charge current, so there is some amount of power dissipated in the IC in order to maintain a regulated output. The LT1516 is an example of a charge pump that uses burst mode to maintain a regulated 5.0 V output. This circuit trades off higher ripple voltage (100 mV at full load) and a second-order filter problem for increased efficiency.

Figure 2-19 shows the internal circuit of the LT1516. Comparator 2 compares the divided output voltage against the internal reference. If the voltage is below the threshold, the charge pump switches are enabled and charge is transferred from the input to the output until the output rises above the upper comparator 2 trip point. This burst mode causes a low-frequency ripple in the output equal to the hysteresis in comparator 2. There is also high-frequency ripple in the output due to the switching of the charge pump while charging the output capacitor.

The LT1516 uses two flying capacitors to implement either a voltage tripling or voltage doubling configuration. Whenever V_{IN} is less than 2.55 V, comparator 1 forces the control logic to put the device in voltage tripler mode. During the charge phase, the switches place both flying capacitors from input to ground. During the discharge phase, the flying capacitor C1 is placed in series with C2 and the series combination is placed in series with the input. Once V_{IN} is

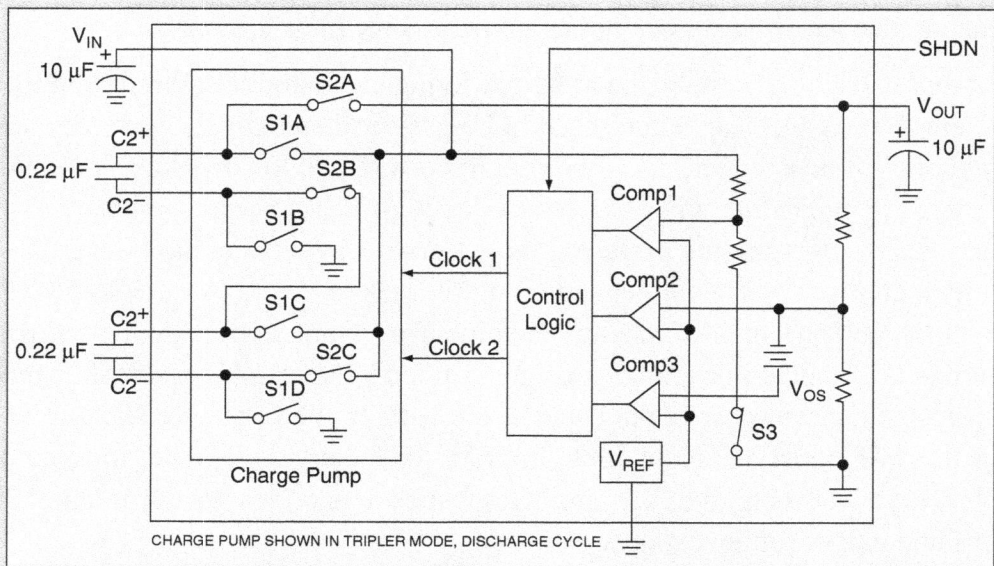

Figure 2-19: Internal circuitry of the LT1516

greater than 2.55 V, the IC switches to voltage doubler mode and just uses C2 as the flying capacitor. Comparator 3 has a 50 mV offset from the feedback voltage at comparator 2. If the voltage droops by 50 mV or more, comparator 3 puts the IC back in tripler mode until the voltage rises above the upper trip point of comparator 3.

The input and output capacitors can be tantalum or electrolytic with the LT1516 because we have comparator control (bang-bang) instead of an error amplifier (proportional control), so there is no control loop to have oscillations. ESR is no longer a consideration for control stability. The only effect ESR will have is on ripple voltage. A good solution is to parallel a low-ESR ceramic capacitor (about 1 µF) with the higher-capacitance electrolytic or tantalum (about 10 µF). The ceramic capacitor reduces 600 kHz ripple from the bursts of charging and the electrolytic reduces ripple at the control frequency.

Multiple Phase PWM Controllers

The demands of power supplies for Pentium® class CPUs are significantly different from low power and traditional CPU needs. A Pentium, Athalon, or Opteron CPU requires very low voltage at tens of amps. A typical Pentium 4 power supply must supply 1.4 V at 65 A.

All of the controllers mentioned so far are single-phase controllers. Current mode PWM controller-based regulators can be operated in parallel to increase current capacity. Many IC manufacturers produce control ICs that operate multiple power supplies in parallel with nonoverlapping phases. The LT3730 is a representative buck controller IC designed for Intelportable computer applications. It can operate up to 600 kHz per phase. Since the phases do not overlap, this gives a ripple frequency of 1.8 MHz. The inductor for each phase can be one-third as large as would be required in a single-phase design. The output capacitor can also be one-third the size of a comparable single-phase design. The input and output capacitor ripple current decreases as you add additional phases. Increasing the number of phases increases the efficiency of the power supply by reducing the losses due to ripple current in the capacitors. Polyphase operation is also possible for boost converters.

Polyphase operation of charge pump controllers can also provide similar improvements in efficiency by effectively increasing the frequency of operation.

Resonant Mode Controllers

One way to increase efficiency and reduce the stress on the switching components is to design the control and filter circuit so that the switch turns on and off at zero current or zero voltage.

Resonant mode switching circuits use constant on-time with variable off-time (as in the TL497) to frequency modulate the current provided to the output circuit. In this case, the inductance and capacitance of the output filter are selected so that the response is resonant at the switching frequency. The frequency is adjusted so that the switch is turned on and off at either zero voltage points or zero current points of the output waveform. The UC1860 is a representative resonant mode switching regulator IC.

Resonant controllers have found very limited use because of the difficulty of design compared with ordinary square wave control. The advantages of resonant mode have been minimized because of advances in MOSFET technology.

CHAPTER 3

The Input Power Supply

- Off-Line Operation
- Radio Interference Suppression
- Safety Agency Issues
- Power Factor Correction
- In-Rush Current
- Hold-Up Time
- Input Rectifier Considerations
- Input Reservoir Capacitor Characteristics

The Input Power Supply

Most of the complexity of the input power supply for an off-line power supply is a consequence of safety and regulatory requirements. We will look at the basics of the AC to DC conversion and then look at the additional circuitry necessary to pass regulatory review.

Off-Line Operation

Power throughout the world comes in a variety of flavors. The major regions are described below:

Region	Voltage	Frequency
USA	117	60
Europe	240	50
Japan	100	60
Middle East	240	50 or 60

Generally, the voltage will vary from ± 10% to ± 15% of nominal. Most commercial power supply designs include Japan requirements with the U.S. requirements and specify the voltage range as either 90–135 or 190–270. A universal power supply must cover the entire range of 90–270 VAC.

Figure 3-1 shows a full wave bridge capacitor input supply for connection to 240 V power. Figure 3-2 shows a full wave voltage doubler supply for connection to 117 V power. Both of these supplies produce a nominal 340 V DC. Figure 3-3 shows a power supply that combines elements of both Figure 3-1 and 3-2 with a switch to produce 340 VDC from either 117 V or 240 V. This circuit is also useful for half bridge circuits, since the capacitor voltage divider is an inherent part of the circuit. These circuits have a worst-case voltage range on the order of 300–410 V.

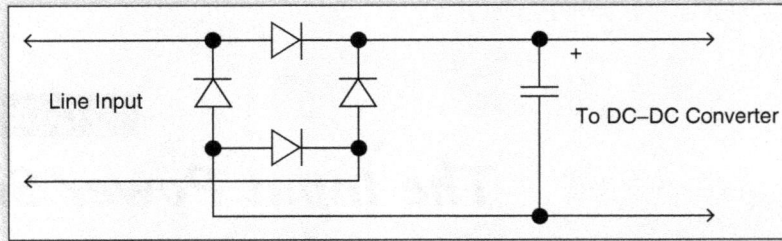

Figure 3-1: Full wave bridge capacitor input supply for connection to 240 V power source

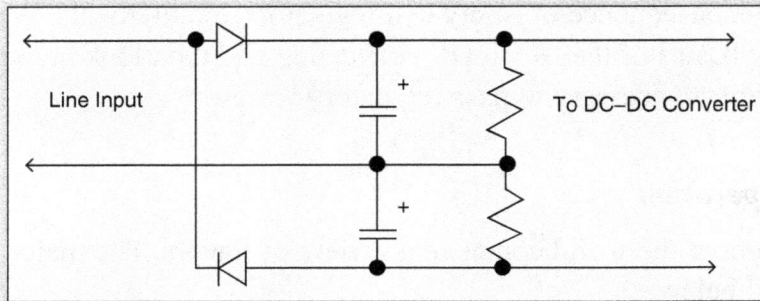

Figure 3-2: Full wave voltage doubler supply for connection to 117 V power source

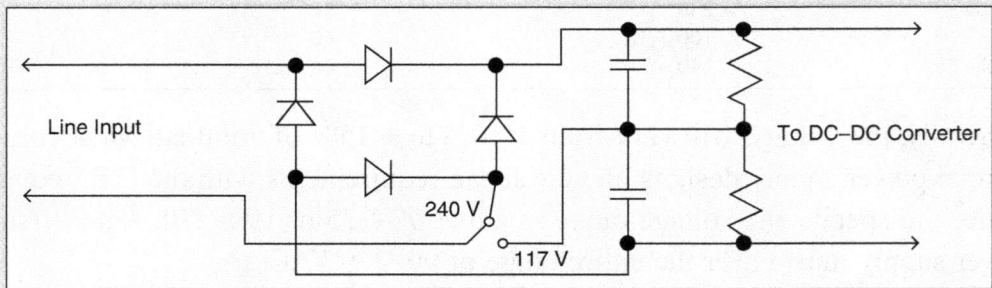

Figure 3-3: Power supply combining elements of Figures 3-1 and 3-2 with a switch to produce 340 VDC from either 117 V or 240 V

The circuit of Figure 3-1 can also be used as a universal input power supply, but the output voltage is no longer a nominal 340 V. This circuit has a worst-case voltage range on the order of 120–410 V. A universal supply will require a much larger range of regulation from the DC–DC converter.

Radio Interference Suppression

Figure 3-4 shows the voltage and current waveforms for the input supply. The bottom trace is the input voltage, the center trace shows the absolute value of the input voltage and the capacitor voltage, and the top trace shows the input current. The rise time of the capacitor current is very short, and so is the fall time. The capacitor current is essentially a very narrow rectangular pulse. This waveform has harmonics that extend up to 5 MHz or higher. Switching current transients from the switch in the DC–DC converter produces noise that is also present at the input of the rectifiers. These transients produce noise at the switching frequency and its harmonics. Even low wattage power supplies can create significant noise that will interfere with radio and television transmissions. Switching power supplies and computers became such a problem in the early 1980s that the U.S. Federal Communications Commission (FCC) created regulation *47 CFR, Part 15 J* that addresses radiation and conduction of digital radio noise outside of equipment. The FCC later changed its regulations to make them equivalent to the Electromagnetic Compatibility (EMC) regulations of the European Community, so a design that will pass in the United States will also pass in Europe. Figure 3-5 shows the input supply and a representative

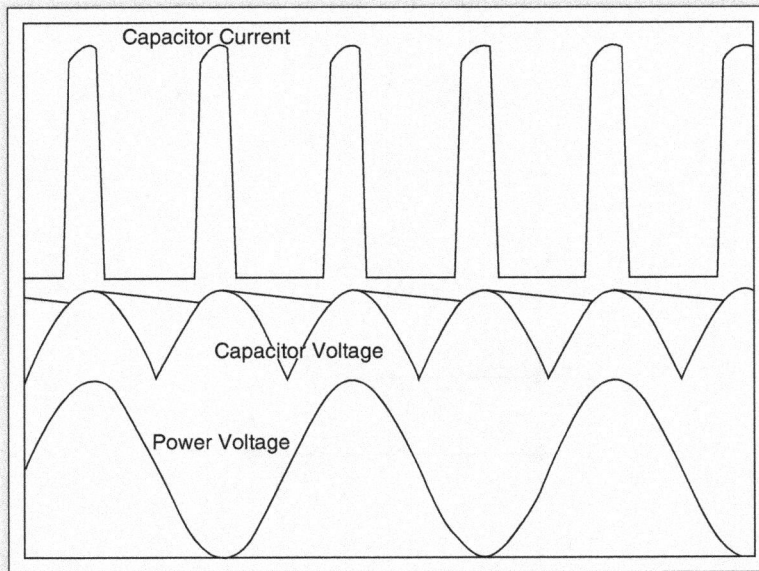

Figure 3-4: Input voltage, capacitor current, and capacitor voltage for the power supply input

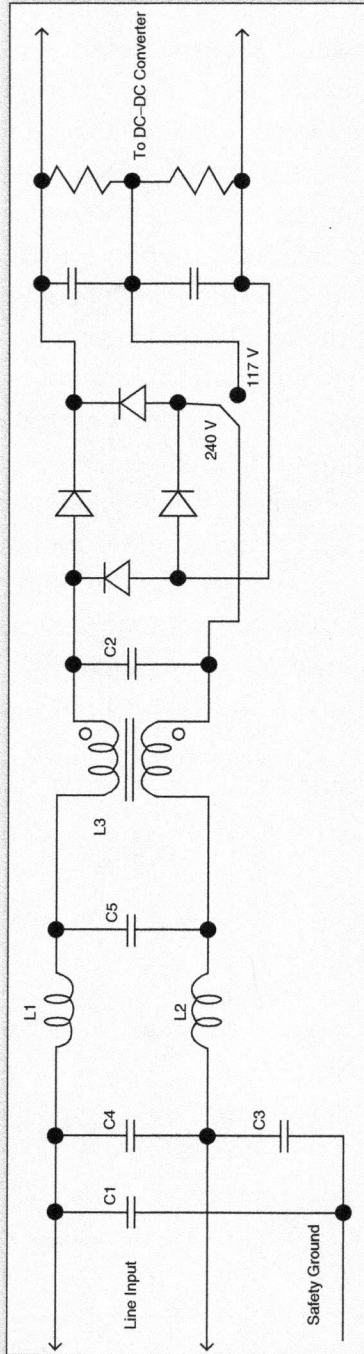

Figure 3-5: Input supply with EMI filter

electromagnetic interference (EMI) suppression filter. This filter has components to suppress conduction of both common mode and differential mode signals. The differential mode signals are a result of direct connections between the switching circuits and the line. The common mode signals are a result of parasitic elements like interwinding capacitance in the isolation transformer or stray magnetic coupling. C4, L1, L2, C2, and C5 are differential mode suppression components. L3, C1, and C3 provide common mode suppression.

Medical devices and devices intended for use where a ground fault circuit interrupter (GFCI) is required must limit the amount of current that flows in the safety ground wire. This imposes stringent requirements on the EMI filter. A GFCI for U.S. style power (with a hot wire, a neutral wire, and a safety ground wire) measures the current in both the hot and the neutral wire. If the currents are not the same, it is possible that a person has created a parasitic circuit from hot to ground and could be electrocuted. The GFCI will trip and disconnect the power line from the outlet. The currents must not be different by more than 1 mA. Patient-area medical equipment must not have current differences greater than 100 μA and should be significantly below 100 μA in a proper design.

It is normal to connect a capacitor between each line and chassis/safety ground. This capacitance provides a leakage current path for power line current to flow. This leakage current will be detected by a GFCI circuit and treated as a hazard condition. In order to control leakage current, capacitors from either hot or neutral to ground should be kept as small as possible. The leakage current limits the size of these capacitors to 470 pF for medical systems and 4700 pF for commercial systems. These capacitors are necessary to limit common mode interference. The circuit of Figure 3-5 will require the EMI filter to be a low-pass filter with a cutoff around 1 kHz.

Safety Agency Issues

Power lines have a multitude of sources of high voltage transients in addition to the normal sine wave voltage. European power producers studied the sources of high voltage transients and the frequency of occurrence. Lightning strikes produce transients as high as 6 kV with duration on the order of 100 ns. The next most important source is a failure in the power network or a failure in equipment located close to the supply (a fuse breaking or flashover in a switch), where the peak voltage is as high as 1.2 kV and duration up to 60 μs. Eighty percent of all transients presented to line operated equipment have duration of 1–10 μs and up to 1.2 kV. The lightning and power network transient voltages are the bases for safety regulations.

In the United States, the appropriate standards for EMI filter capacitors are UL1414, UL1283, and IEC 950 [as adopted by Underwriters Laboratories, Inc. (UL)]. Canadian standard CSA C22.2 No.1 is equivalent to UL1414 and CSA C22.2 No. 8 is equivalent to UL1283. These standards are only applied to radio, television, and certain telecommunications equipment. The standard used in the European Community is EN132400 (previously IEC 384-14). This standard is significantly more comprehensive and stringent than the North American standards. Designing for the European standard should be adequate, in most cases, to also meet the North American standards.

EN132400 defines seven classes of capacitors for use in EMI filters in line operated equipment. Currently, only X1, X2, Y1, and Y2 ratings are applied to equipment classes. Class X capacitors are connected from line to line in 220 V U.S. and European systems, and from line to neutral in 110 V systems. Class Y capacitors are used from either line to safety ground. Class Y2 capacitors are the most common type of Y capacitor and are used in systems such as computer power supplies. Class Y1 capacitors have more stringent requirements since they are intended to connect either line to ground in double-insulated equipment. Class X1 capacitors are specified for use in equipment that is permanently attached to the power system of a building, such as a mainframe computer or lighting ballast. Class X2 capacitors are the most common type of X capacitor and are specified for use in equipment plugged into wall outlets.

One of the most common failure modes of a capacitor during a voltage transient is a short circuit through the dielectric. Metallized paper and metallized film capacitors are designed to self-heal after a transient-induced failure. The current level in the capacitor is very high at the point of failure. The high current level melts the metallization and moves it away from the hole in the dielectric.

This area is isolated from the rest of the capacitor and can no longer fail. The capacitor remains functional after it heals.

The self-healing process may leave a conductive residue behind. If enough self-healing events occur during the life of a capacitor, it may lead to excessive current flow due to conductive residue. If enough conductive residue builds up, the capacitor will fail due to heating and excessive leakage current. Excessive heating can lead to equipment fires.

Metallized paper and metallized film are the two types of capacitors that have the least chance of creating resistive residue during self-healing. The amount of free carbon in the dielectric is a direct indicator of the likelihood of creating resistive residue. Paper and polyester dielectrics have the smallest amount of free carbon and are the preferred materials for EMI capacitors.

Ceramic capacitors do not self-heal, so they must be manufactured with enough dielectric strength to withstand the applied transients. Ceramic capacitors can fail with short circuits. This requires that ceramic RFI capacitors are manufactured in larger physical sizes.

Y capacitors must be limited to a maximum value that is dependent on the permissible leakage current. We must choose a nominal capacitor value that will guarantee that we do not exceed the maximum leakage with environmental changes. There are several factors that change the value of a capacitor: temperature coefficient, aging, voltage dependence, and initial tolerance.

X2 capacitors are required to withstand 2.5 kV transients and Y2 capacitors must withstand 5 kV transients. Y capacitors must withstand higher voltage since a failure where the leakage increases will increase the electric shock potential. A failure of an X capacitor will cause the device to fail, but will not increase the risk to the operator as long as it does not catch fire. Flammability is another measure of suitability of X and Y capacitors.

Safety agencies are only concerned that a failure does not cause danger for an operator. IEEE standard 587 addresses additional parameters intended to ensure that a system will not fail in the presence of power line transients. The most likely cause of failure of a system is a lightning-induced transient. This is why IEEE 587 specifies testing with a damped sinusoid with 6 kV peak amplitude.

Power Factor Correction

Power factor is defined as the ratio of the actual power consumed by the device divided by the apparent power consumed by a device. A device with a power factor of 1.0 has a sinusoidal current that is identical in phase and frequency to the applied voltage. In this case, the apparent power (magnitude of current times magnitude of the voltage) is the same as the real power. If the current is out of phase from the voltage, the apparent power can be considerably higher than the actual power consumed. For linear loads such as resistive heaters or motors, power factor is calculated as cosine(Φ), where Φ is the phase difference between current and voltage. Zero degrees gives a power factor of 1.0; 45 degrees gives a power factor of 0.707; and 90 degrees gives a power factor of zero. Power factor for nonlinear loads such as power supplies requires more sophisticated measurement techniques because the current waveform is not sinusoidal.

A power factor of zero has serious consequences for the power company. The actual power consumed is zero, but the current supplied by the power lines is still equal to the amount of current flowing in the device. Losses in the wires delivering power to the consumer are I^2R, regardless of the power consumed, so a consumer could cause losses in the transmission system while actually not consuming any power. Power throughout the world is transmitted using three phases, each 120 degrees from the next. The third harmonic created by the short current pulses in Figure 3-4 is especially problematic for three-phase systems because they add in phase in the neutral wire. It is possible for the third harmonic current to exceed the current capability of an otherwise properly sized neutral wire. The European Community has passed regulations (IEC 555, EN61000-3-2) that have the effect of requiring most systems to have power factor correction. These regulations actually restrict the size of the harmonics that are presented to the power distribution system, but power factor correction is an effective means to meet the regulations.

You can use either active or passive means to increase the power factor. Figure 3-6 shows two mechanisms to increase power factor using passive circuits. The easiest passive means is to use a choke input filter rather than a capacitor input filter for the power supply. The current drawn from the power line still occurs

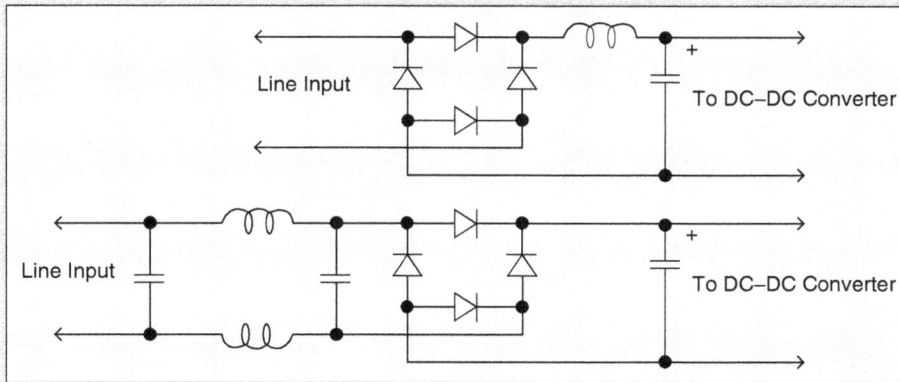

Figure 3-6: Methods to increase power factor through the use of passive circuits

in pulses that are shorter than the input voltage, but the rise and fall times of the current are much longer. A choke input filter will also conduct for a longer time during the input waveform. The current is essentially sinusoidal-shaped pulses. Increasing the rise and fall time reduces the energy at higher harmonics. Additional improvement in power factor can be obtained by setting the cutoff of the EMI low pass filter as low as practical to reduce harmonics of the power frequency. This will require multiple inductors and capacitors. Passive correction requires rather large inductors for both the choke input and the power line filter, since they are acting at frequencies on the order of 100 Hz. This is problematic if the size of the power supply is important. Another issue for the choke input supply is that it is only effective where the inductor current is continuous.

Figure 3-7 shows a typical means of implementing active power factor correction. This circuit uses a boost converter running at a high frequency (typically 100 kHz) with an EMC filter to reduce conduction of 100 kHz switching transients back into the power line. The power factor control (PFC) IC adjusts the amount of current drawn from the line so that it is a constant factor of the input voltage. This makes the input of the boost converter appear to be a resistive load. The output voltage of the boost converter must be greater than the peak input voltage to guarantee that the inductor always delivers its energy to the output capacitor. The control circuitry is more complicated than an ordinary boost converter. An ordinary boost converter must only adjust the duty cycle to

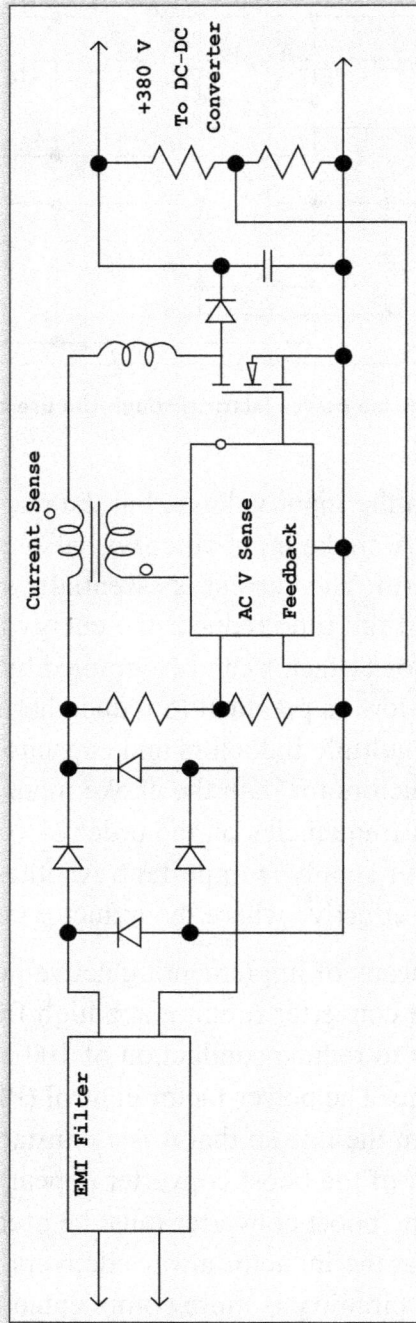

Figure 3-7: Active power factor correction through the use of a high frequency boost converter

produce the correct output voltage from the input voltage. The PFC circuit must also adjust the duty cycle during each sample period so that the current stays in phase with the input voltage. The voltage out of the PFC circuit does not need tight regulation, since the PWM DC–DC converter will provide the final output regulation. In fact, the limited regulation provided by the PFC makes the design of the PWM section significantly easier.

There are numerous issues that must be addressed in an active PFC/PWM circuit because it is composed of two servo systems in series. Several manufacturers make ICs that combine the PFC and PWM circuits to make it easier for the designer to avoid the pitfalls of active PFC. Another advantage of the active PFC circuit is that the EMI filter need only attenuate the switching frequency and its harmonics, since power frequency harmonics are minimized as part of power factor correction.

Figure 3-8 is from the data sheet for the LT1248 PFC controller and shows the internal components of the IC and external circuitry to implement active power factor correction. Most of the circuitry looks like a standard voltage mode boost PWM controller. It has a voltage reference, soft start circuit, undervoltage lockout, and an oscillator that creates a ramp that is compared to the command signal to adjust the duty cycle. However, the circuit that creates the command signal is quite different.

The PWM control circuit measures input line voltage, input line current, and output voltage in order to set the duty cycle. The first circuit in the control chain is an error amplifier that measures the output voltage. It is a standard PWM error amplifier with the output available for compensation. The output of the error amplifier is converted into input current for the input voltage multiplier. The input line voltage is converted into a current that is proportional to the input voltage and applied to the second input of the multiplier. The output of the multiplier is applied to the noninverting input of the current amplifier. The current amplifier measures the voltage across the current sense resistor. The multiplier current is summed with the current sense to adjust the control voltage. Again, the output of the current amplifier is brought out to a pin so that this amplifier can be individually compensated. This compensation is necessary to force the current amplifier to respond to double the line frequency (it is double because of the full wave rectification) rather than the switching frequency. The multiplier in this part is unusual in that it adjusts its response as a square of the error amplifier input. The error amplifier signal is a response to the output load. As the load is reduced, the signal from the current sense resistor will drop. The square response of the multiplier helps to keep the gain of the current control adjusted to give better stability.

All PFC controllers need to implement over-voltage protection for the output. The control loop is designed primarily to have the line current follow the input voltage, so the transient response is quite slow. If the load current drops significantly and quickly, the output voltage can overshoot because of the current that is flowing in the boost inductor. This phenomenon is called "load dump" because the inductor cannot dump its energy into the load. The over-voltage protection circuit immediately shuts off the switch and bypasses the control loop to handle this transient condition.

In-Rush Current

The power switch can apply power to the input rectifier at any point on the input AC waveform. The reservoir capacitor will always look like a short circuit when power is first applied. If the power switch is turned on when the input voltage is at its peak value, a very large current results that is limited only by the resistance of the EMI filter and any other resistances ahead of the rectifier. This in-rush current can be on the order of 20–1000 times the normal peak current in the system. In-rush current can be destructive to the capacitors, the rectifiers, and the power switch. Figure 3-9 shows three ways to add series resistance to the system during the initial charging of the reservoir capacitor. The triac and SCR are most appropriate in high power supplies where the additional cost is a minor portion of the total. The triac is a poor choice for a PFC input because it would add more power frequency harmonics. The negative temperature coefficient thermistor is a reasonable method that is chosen by most designers. Manufacturers produce NTC thermistors specifically designed for in-rush current limiting where the ratio of resistance at room temperature to operating temperature is quite high. NTC thermistors do not provide protection

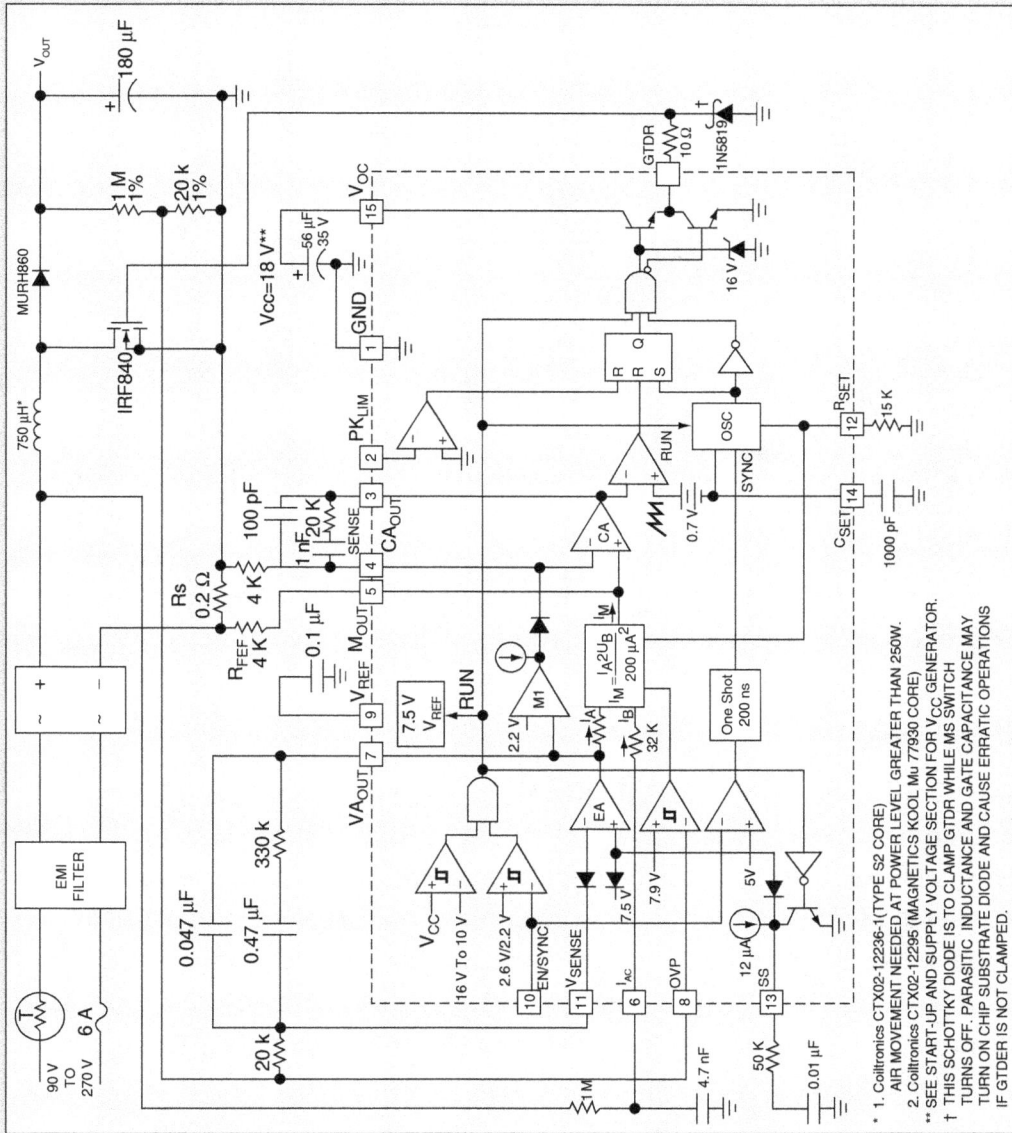

Figure 3-8: Internal circuitry of the LT1248 power factor control (PFC) controller

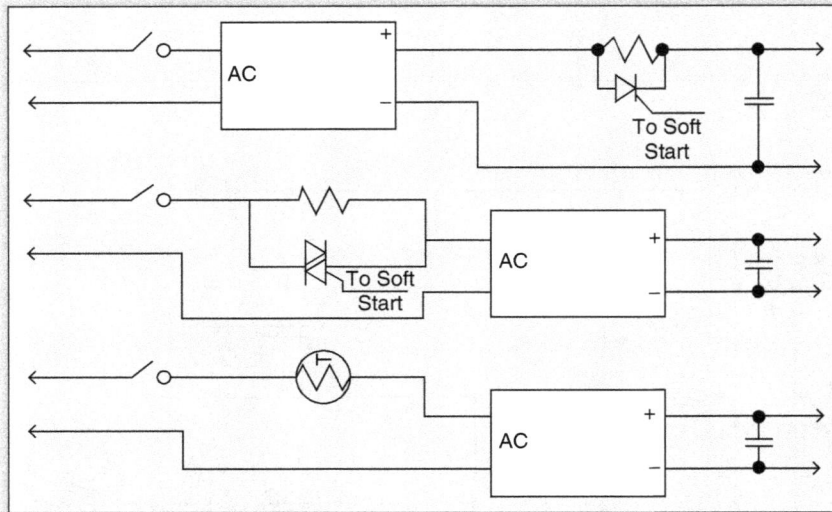

Figure 3-9: Three methods to add series resistance to the system during the initial charging of the reservoir capacitor

if the power is cycled multiple times in a short period. The thermistor must cool off between power cycles in order to provide full protection.

Hold-Up Time

As described above, power lines do not always provide a clean, constant sine wave. Frequently, the power line will drop out for one or more half-cycles. Many systems, such as mission-critical computer systems, cannot tolerate an uncontrolled power cycle. A switching power supply for such a system must be designed with a large enough reservoir capacitor and enough range in the PWM circuit to continue supplying full power during loss of multiple cycles. We have two options for supplying load power during power main failure. The first is to increase the storage of energy on each output by using large output capacitors, and the second is to increase the storage of energy in the input supply.

There are several reasons why increasing the output capacitors is seldom the selected method. The first problem is that each output capacitor in a multiple output supply must be increased. This problem is compounded by the amount of additional capacitance required. The energy stored in a capacitor is $\frac{1}{2} CV^2$, so it requires considerably less capacitance to store energy on the input side where the voltage is 340 V rather than where the voltage is 5 V or 12 V. Larger output capacitors can also cause longer transient response. It follows from the analysis above that we would choose a voltage doubler supply rather than a universal full-wave bridge in any situation where the power voltage is 117 V and we require a long hold-up time. A voltage doubler supply will increase energy storage by four times for the same amount of capacitance.

Figure 3-10 illustrates one of the problems with a normal capacitor input filter when the power line drops cycles. The capacitor is only charged during about 20% of the half-cycle at the peak of the input waveform. If the power line should drop out immediately before the capacitor would begin charging, as in Figure 3-10, we have to add essentially one complete half-cycle to the amount of time that the

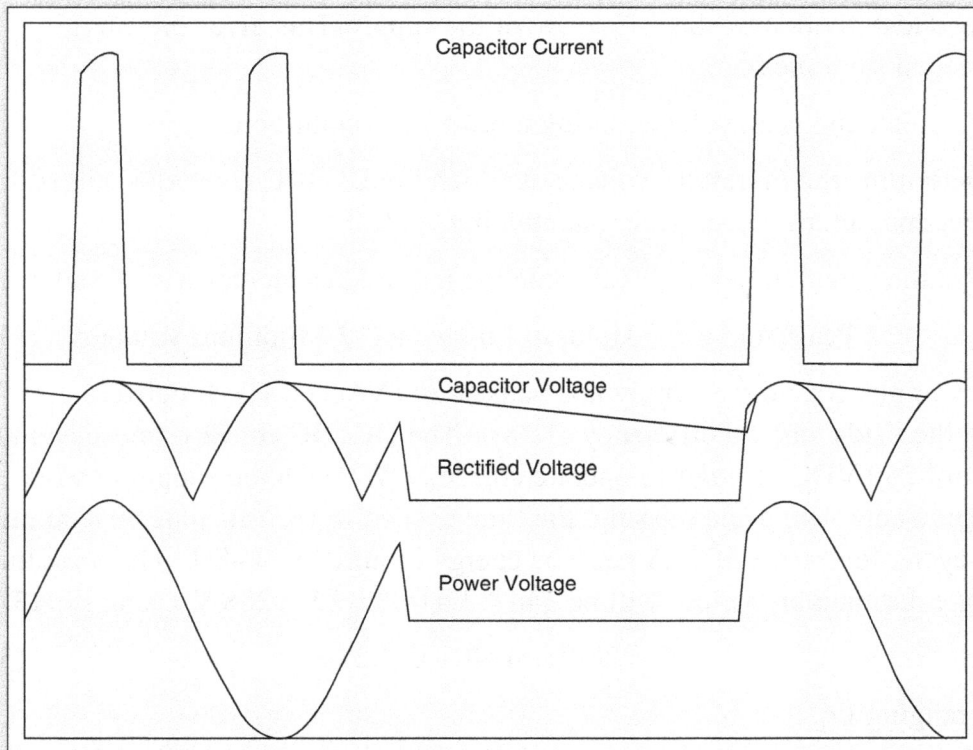

Figure 3-10: Worst-case capacitor voltage, capacitor current, input voltage, and rectified voltage during lost input cycles

power is off for our worst-case analysis. For a 50 Hz system, this adds approximately 8 ms to the minimum hold-up time for the supply. Another issue for a capacitor input filter is the line voltage when the power drops out. If the power drops out at the same time that a brown-out is occurring (a normal situation during a thunderstorm), then the energy stored in the capacitor is at its lowest. Our calculations of minimum capacitance must now include low voltage as well.

Here is an algorithm for calculating the minimum capacitance required for the input reservoir:

1. Calculate the time that the supply must work with no input power. This is the number of half-cycles times the period of the power frequency plus 80% of one half-cycle.

2. Calculate the energy that will need to be delivered while the power line is dead. This is simply maximum power output (in watts) times the time (in seconds) divided by the efficiency of the supply. This gives the energy needed (in joules).

3. Calculate the peak voltage during a brown-out condition.

4. Determine the minimum voltage at which the DC–DC converter will still give maximum power while maintaining control.

5. Use the equation below to calculate the required capacitance (in farads).

$$C * \text{Peak Voltage}^2 = \text{Hold-up Energy} + C * \text{Minimum Voltage}^2 \qquad (3.1)$$

For example, consider a supply that runs at 240 VAC 60 Hz. It delivers 150 W to all the loads with an efficiency of 78%. The DC–DC converter needs a minimum of 250 VDC in order to operate correctly. We wish the supply to work with one complete cycle missing, the time needed is 16.7 ms plus 80% of one half-cycle, for a total of 23.3 ms. The energy required is 4.48 J. The peak line voltage during a brown-out will be 240 * 1.414 * 0.85 = 288 V. Using Eq. (3.1),

$$C * 288^2 = 4.48 + C * 250^2 \qquad (3.2)$$

Rearranging Eq. 3.2,

$$82944 \, C - 62500 \, C = 4.48 \qquad (3.3)$$

$$C = 4.48/20444 = 220 \, \mu F. \qquad (3.4)$$

Active power factor correction has an added benefit of improving hold-up time. An active PFC circuit keeps the output voltage relatively constant, so the hold-up time is no longer dependent on where in the cycle the line voltage drops out. Energy is lost only for the time period when the power is actually dead. The output voltage is also kept fairly constant with changing line voltage, so the hold-up time is not dependent on the line voltage.

Input Rectifier Considerations

Linear power supplies use an iron core transformer to isolate the circuitry from the power line and transform the voltage to an appropriate value. This transformer blocks many of the negative effects of the transients that are present on the power line. The primary mechanism of transmission of transients to the secondary circuit is the capacitive coupling between the windings. This isolation reduces the stresses presented to the rectifiers.

The rectifiers in off-line switching supplies are subjected directly to transients on the line, so they can be subjected to very large voltage pulses and very large surge currents. In order to protect the input rectifiers, it is advisable to include a transient protection device on the input line. There are only two types of devices that are practical for transient protection: zener diodes and metal oxide varistors (MOVs). Zener diodes can be used back-to-back, but high voltage zener diodes tend to be expensive. Vishay Semiconductors makes a line of zener diodes called TransZorb designed specifically for transient suppression. MOVs are avalanche devices that have a very high resistance until they reach the avalanche voltage. When the avalanche voltage is reached, they short and absorb the energy of the transient. After the voltage goes below the avalanche point, they revert to a high resistance. The voltage rating of the MOV must be selected to be significantly above the highest expected line voltage. If the input power voltage should exceed the avalanche voltage, the MOV will attempt to clamp and it will immediately fail. Failure of an MOV is usually a fracture. If the failure is due to power voltage, the MOV will likely explode. It is best to include some sort of mechanical protection in case the MOV fails catastrophically. Although the MOV or zener diodes will limit the peak voltage of transients, it is best to choose 1000 V PRV diodes to allow adequate surge voltage protection.

The input diodes only conduct for a very short period of the total input voltage. The rectifier current can be as much as 10 or 20 times the RMS line current. The size of the reservoir capacitor has a direct impact on the ratio of peak to average current. Using a larger capacitor for a longer hold-up time will increase the ratio and this will require rectifiers with higher average current ratings. A good rule of thumb is to limit the peak-to-average ratio to 20, as an absolute maximum.

It is also important to choose rectifiers that have adequate power dissipation. The Motorola MDA970A6 bridge rectifier has a 4.0 A average current rating at 20°C, but it drops to 2.0 A at 80°C. You must also consider the power dissipation and thermal resistance to evaluate the junction temperature. We learned a rule of thumb for low current applications is that the diode voltage drop is 0.7 V. The diode voltage is an exponential function of current. At 10 A of current, the voltage drop is over 1.0 V for silicon diodes. This value is usually given on the data sheet and can be used with average current flow to determine power dissipation.

Input Reservoir Capacitor Characteristics

Aluminum electrolytic capacitors are the only variety that will give sufficient voltage capability and high capacitance in a reasonable physical size. A single capacitor with 400–450 V rating is adequate for 240 V operation. For a voltage doubler in a 117 V system, two capacitors with 200–250 V rating will be adequate. Such a system will require high value bleeder resistors in order to equalize the voltage on the capacitors. Without equalizing resistors, one of the capacitors can have significantly higher voltage than the other, which could lead to over-voltage and destruction of the capacitors. In order to ensure that there is enough margin, it might be necessary to choose capacitors with 300 V or more rating.

The capacitor current is an AC waveform with a large, very short positive value during charging and a longer discharge. As a first approximation, the RMS value of the AC current is equal to the DC current provided to the DC–DC converter. This AC current causes heating due to power dissipation in the equivalent series resistance (ESR). Not all electrolytics are rated for such high AC current service. It is important to choose a capacitor that is rated for high ripple current service. See Chapter 6 for detailed descriptions of the parameters of electrolytic capacitors.

CHAPTER 4

Non-Isolated Circuits

- General Design Method
- Buck Converter Designs
- Boost Converter Designs
- Inverting Designs
- Step Up/Step Down (Buck/Boost) Designs
- Charge Pump Designs
- Layout Considerations

Non-Isolated Circuits

In this chapter, we will look at detailed designs of non-isolated converters. The applications include remote regulation for line-operated systems or power management in battery-operated systems. I find at least one new application or new device suitable for a non-isolated circuit every week in the trade journals. The applications have exploded during the past five years and show no signs of slowing. The trend is toward ever smaller, more efficient, and more specialized controllers. The designs we will look at here will give general design methodology as part of the specific designs.

Engineers seem enamored of creating new jargon and acronyms or initials. The new term for remote regulation is point of load (POL). Point of load regulators are almost always non-isolated circuits.

All of the designs shown here use current mode PWM control because of its inherent advantages in loop stability and current control. One of the problems with current mode control is subharmonic oscillation at duty cycles greater than 50%. Older ICs required external means to provide slope compensation to eliminate subharmonic oscillations. The modern ICs described in this chapter all contain internal slope compensation, so there is one less design task to complete.

General Design Method

There are many design variables for power supply designs, so each design will be different from the last one. The sequence below will help beginners with a starting framework. A complete design will usually require iterations of several steps.

1. Choose a converter type based on the input voltage range and the output voltage. Input always above the output indicates a buck converter. Output voltage always above the input indicates a boost design.

2. Choose an IC based on the output power, physical size, etc. This is probably the most daunting task for a beginner because there are so many parts from so many vendors. The complexity of the circuit is usually dictated by the output power. The more power, the larger and more complex the circuit. The requirements will normally force the selection of switching frequency at this step. We usually decide on diode or synchronous rectification here.

3. Choose the ripple current in the inductor based on the output ripple voltage requirement. This decision affects the choice of the input and output capacitor because of the interaction of capacitor ESR and ripple current.

4. Calculate the inductor value based on the ripple and average current.

5. Calculate the required current sense resistor based on the IC data sheet.

6. Select the switch transistor and diode based on the inductor current.

7. Calculate the input and output capacitor values based on the ripple current and ripple voltage requirements.

8. Select a first try at the loop compensation circuit.

9. Select the soft start components, if required.

Buck Converter Designs

The LT1765 is a full function current mode PWM IC with an integral NPN transistor switch, current sense resistor, and slope compensation. The switch frequency is fixed at 1.25 MHz. Figure 4-1 shows a representative buck converter design using the LT1765. This part is available in either an SO8 or 16-pin TSSOP package. The SO8 package uses the lead frame connected to the ground pin for heat dissipation. The TSSOP package has an integral heat sink pad under the package to conduct heat to the ground plane. This part is designed for small size and low bill of materials.

Any buck converter that uses an NPN transistor or NMOS switch will require a voltage above the input voltage in order to fully turn on the switch. A bipolar

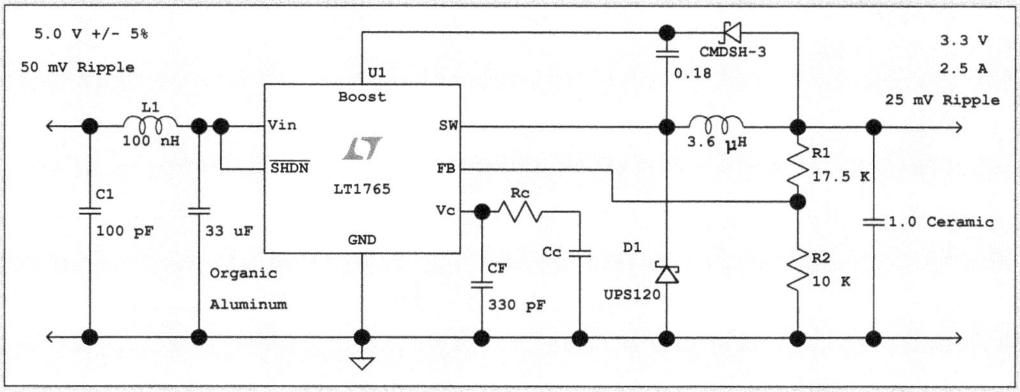

Figure 4-1: Representative buck converter using the LT1765

switch will only need a control voltage that is 0.7 V larger than the input voltage. The control voltage for an NMOS switch will be higher than for a bipolar switch. If an NMOS switch is used, the best choice for a buck converter is a logic-level switch that will only require about 2 V above the input voltage (see Chapter 7 for details on switch parameters).

Figure 4-1 shows a charge pump implementation that supplies the necessary switch control voltage. This concept will work for both bipolar and NMOS designs. When the switch is closed, the voltage of the boost capacitor will add to the switch voltage so that the switch can saturate. When the switch opens, the boost capacitor will be connected across the output and charged to the output voltage minus the two diode drops from D1 and D2 (about 1 V less than the output voltage). The diode voltage drops and the internal supply circuit voltage drop limit the output voltage for full efficiency to about 3.3 V. If a lower output voltage is required, the switch will no longer saturate and power dissipation will increase dramatically. The data sheet for the LT1765 recommends a 0.18 µF boost capacitor for most applications. This value is calculated based on 700-ns on time (87% duty cycle), 90 mA boost current, and 0.7 V of ripple on the boost voltage. A ceramic capacitor with ESR below 1 Ω will be required to fully charge the capacitor during the shortest off time.

This circuit will bootstrap the boost voltage during powerup. When the circuit starts, the output voltage and the voltage at the switch pin will be zero. The control circuit will turn on the switch and the switch pin voltage will be 0.6 V below the input voltage because of V_{BE}. The transistor will not be saturated, but it will begin supplying inductor current and begin charging the output capacitor. As the output voltage rises above 1.0 V, the boost diode will conduct when the switch is off and begin charging the boost capacitor. The power dissipation in the switch will rapidly decrease as the boost voltage increases.

Current mode PWM controllers provide inherent output current limiting in a buck converter. The output current will be limited to the peak inductor current. For PWM ICs that have a shutdown pin, you can use external circuitry to detect a fault and shut down the power supply.

The size of the inductor determines the amount of ripple current. We use the inductor equation and the duty cycle equation to determine the relationship between the inductor and the ripple current.

Equation (1-6) from Chapter 1 gives the duty cycle in terms of the voltages:

$$V_O = V_{IN} * DC, \text{ or } DC = V_O/V_{IN}$$

Equation (1-1) from Chapter 1 gives the inductor voltage in terms of inductance and change in current:

$$V = L * (\Delta I/\Delta t)$$

The amount of time for the current to go from minimum to maximum is:

$$\Delta t = T * DC, \text{ or } \Delta t = (1/f) * DC \text{ or } \Delta t = (1/f) * (V_O/V_{IN})$$

where T is the period of the switching frequency f.

We can rearrange the inductor equation to yield:

$$L = V (\Delta t/\Delta I), \text{ or } L = (V_{IN}-V_O) * (\Delta t/\Delta I), \text{ or }$$

$$L = (V_{IN}- V_O) * (V_O /(\Delta I * f * V_{IN}) \qquad (4\text{-}1)$$

One of the parameters that affects the design is the range of the input voltage. The ripple current is greatest at the highest input voltage. A good rule of thumb is to set the ripple current equal to 10% of the maximum output current at the highest input voltage. We do not have control over the maximum switch current because it is set to 3 A by the circuitry of the IC. The maximum available output current will be 3A − $\Delta I/2$ − 70 mA (boost current).

Using our rule of thumb, we will set the ripple current to 250 mA. We can plug the values into Eq. (4-1):

$$L = (5.0 - 3.3) * (3.3/(0.25 * 1.25 * 10^6 * 5.0) = 3.6 \ \mu H$$

Transient response and ripple current are related. A large ripple current will allow faster response to load changes. However, large ripple current combined with the ESR of the output capacitor will increase the output ripple voltage. Figure 4-2(a) shows the equivalent AC circuit for the output when the output capacitor is infinite. If $(10 * ESR)$ is less than the value of R_L, then we can make the simplifying assumption that all of the ripple current flows in the ESR of the capacitor. If we consider the capacitor leg to be ESR in series with the capacitive reactance, as in Figure 4-2(b), then we can use this impedance to set the output ripple voltage.

Ripple voltage is usually set as a design parameter, so we can use it to select the size of the capacitor and its ESR.

The peak-to-peak ripple voltage is found by:

$$\Delta V = \Delta I * (ESR + X_C)$$

Substituting and rearranging, we get:

$$ESR + X_C = \Delta V * (L * f * V_{IN})/(V_O *(V_{IN} - V_O))$$

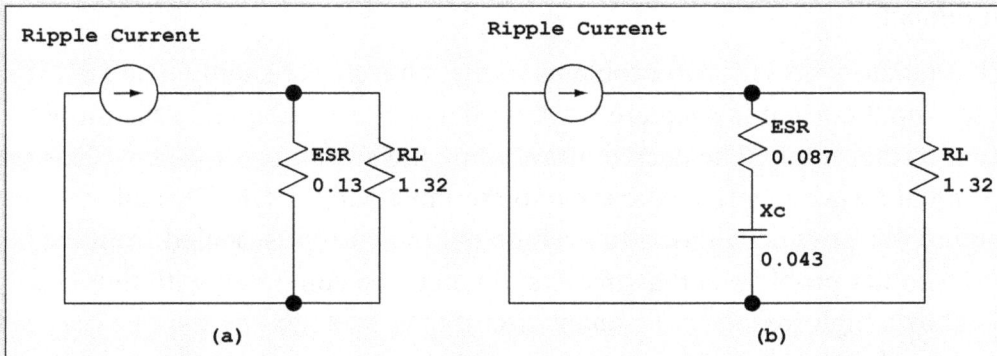

Figure 4-2: (a) Equivalent AC circuit for the output when the output capacitor is infinite; (b) equivalent AC circuit with ESR in series with the capacitive reactance

A good rule of thumb is to allocate two-thirds of the total impedance to the capacitor ESR and the remaining one-third to the capacitor. We can use the formula for capacitive reactance to decide on a capacitance value:

$$C = 1/(2 * \pi * f * X_C)$$

This value will give a slightly larger value than necessary because the waveform is triangular rather than a sine wave and the higher harmonics will be attenuated to a greater extent. The consequence of the decision to allocate one-third of the impedance to the capacitor is that as ESR goes down, we can use smaller capacitor values. It is possible that a capacitor with the required capacitance will have ESR that is larger than the target ESR, especially for aluminum electrolytic capacitors. If this is the case, then it will be necessary to increase the capacitance or allocate more of the ripple budget to the ESR of the capacitor. Obtaining reasonable transient response, ripple voltage, and loop stability may require several iterations to obtain a design that meets all the criteria.

The specification in Figure 4-1 calls for 25 mV of ripple. Using the equation above,

$$ESR + X_C = 0.025 \text{ V} * (3.6 \text{ } \mu H * 1.25 \text{ MHz} * 5.0 \text{ V})/(3.3 * (5.0 - 3.3)) = 0.100 \text{ } \Omega$$

We are looking for a capacitor with 0.07 Ω ESR and 0.03 Ω reactance. This calculates to 4.3 μF. A multilayer ceramic capacitor is a reasonable choice for this capacitor. Since ceramic capacitors have almost no ESR, a capacitor between 1.4 μF and 4.3 μF will be likely to satisfy our output ripple requirement.

Buck regulators present two problems to the input power supply. The first is that the input current is a square wave with a peak value equal to the output current of the supply. The current draw while the switch is off is zero. This very large square wave is reflected back into the input supply. L1, C1, and the 33 μF of Figure 4-1 provide filtering to average out the current supplied from the input. Another problem is that any stray inductance combined with the 33 μF will act as a high frequency resonant circuit that is excited by the fast rise and fall times of the current. This can cause EMI problems at harmonics of the switch frequency. See the section on Layout Considerations later in this chapter for more details.

The RMS ripple current in the input capacitor is determined by:

$$I_{RMS} = I_{OUT} (DC - DC^2)^{1/2} \qquad (4\text{-}2)$$

It is important to choose a capacitor that is rated for this ripple current. The RMS input ripple current is 1.2 A. We are given a budget of 50 mV input ripple voltage, so the capacitor impedance must be 0.04 Ω or lower. A Kemet 33 μF organic aluminum electrolytic will have 0.028 Ω ESR and will handle 2.1 A ripple current with 8 WV or 10 WV.

L1 and C1 are optional input filter components that will improve EMI performance of the supply. The input filter components can have an adverse impact on loop stability. Buck regulators have a negative resistance characteristic for low frequencies. As the input voltage drops, the input current rises to maintain the output voltage. If the input filter has a high Q, it is possible for the negative resistance of the buck regulator to combine with the input filter to produce a sine wave oscillator. This is another place where you must balance competing goals. The attenuation characteristics of the filter must be balanced with stability. Lowering the resonant frequency will increase attenuation, but can lead to loop instability. This is a place where iteration in the lab is likely to be necessary to obtain a stable power supply.

The data sheet gives us guidance on compensating the feedback loop. We will start with 330 pF for C_C, and 0 for R_C and C_F. If we were to build this design, we would adjust the values of these three components in the lab to account for second-order effects of the components and the effects of the circuit layout, using the compensation method described in Chapter 1.

The data sheet also gives us guidance on selecting R1 and R2. Linear Technology suggests 10 k for R2 to minimize the offset voltage due to bias current of the feedback pin. The formula for R1 is given as:

$$R1 = \frac{R2 \times (V_{OUT} - 1.2)}{1.2 - (R2 \times 0.25\ \mu A)} = 17.5\ K$$

Figure 4-3 is a circuit from the data sheet that gives a soft start circuit using external components connected to the compensation pin. This circuit can be used for any current mode PWM controller that does not provide an internal

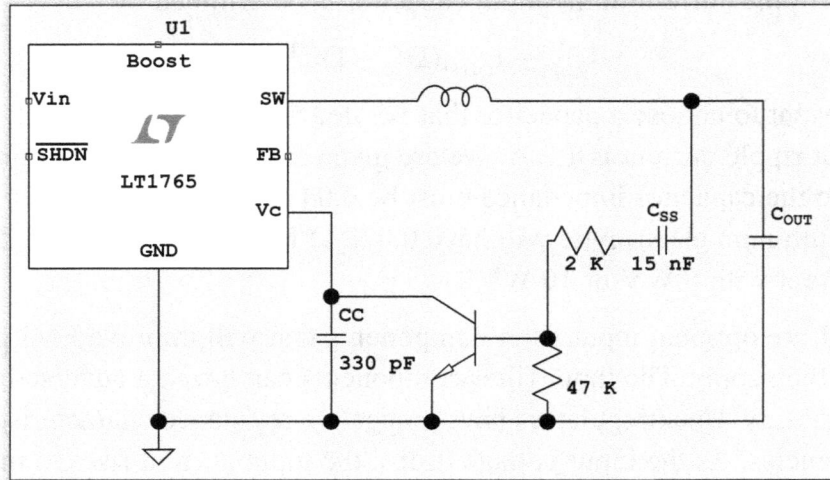

Figure 4-3: Soft start circuit using external components connected to the compensation pin

soft start circuit. The soft start works by limiting the voltage rise on the compensation pin. The circuit effectively adds the soft start capacitor (C_{SS}) to the compensation capacitor to create a very heavily damped response. As the output approaches the final value, the extra damping gradually decreases so that only the 330 pF controls compensation.

Diode D1 of Figure 4-1 will have a forward voltage drop of 0.4 V at 3 A current. The equations up to this point have made the simplifying assumption that the diode forward drop is so small that it can be ignored. In the case of Figure 4-1, this is arguably not valid. As long as the input voltage is well regulated, the errors do not affect the final outcome; the circuit will still be able to maintain control. However, if the input voltage has a larger range, the circuit may have more trouble maintaining control. We need to add the diode drop to the output voltage in each of the equations where V_O appears as the voltage across the inductor in order to get accurate results.

DC becomes:

$$DC = (V_O + V_D)/V_{IN},$$

so DC = (3.3 + 0.4)/5.0 = 0.74 instead of 0.66.

This change in duty cycle will affect the value of ripple voltage, ripple current, and the value of the inductor. Equation (4-3) gives a more accurate equation for the inductor value:

$$L = (V_{IN} - V_O) * (V_O + V_D)/(\Delta I * f * V_{IN}) \tag{4-3}$$

For Figure 4-1, this changes the inductor from 3.6 µH to 4.0 µH.

The average output diode current can be found from:

$$I_{AVG} = I_{OUT} * (1 - DC)$$

The power dissipation in the diode at full load in Figure 4-1 is $(2.5 * (1 - 0.74) * 0.4) = 0.26$ W. We also need to account for losses in the switch in the IC. The worst-case saturation voltage is 0.43 V. The average switch current is:

$$I_{AVG} = I_{OUT} * DC$$

The switch power in Figure 4-1 is $(2.5 * 0.74 * 0.43) = 0.80$ W. The actual power dissipated in the switch is slightly higher due to the slope of the switching waveform. The data sheet gives this value as:

$$17 \text{ ns} * I_{OUT} * V_{IN} * f$$

This gives total switch loss of 0.80 W + 0.27 W = 1.1 W. The boost circuit also dissipates power. The data sheet gives a formula for boost circuit dissipation:

$$P_{Boost} = (V_O^2 * (I_{OUT}/50)/V_{IN}) = 0.1 \text{ W}.$$

The boost circuit draws 70 mA during switch on time, so this power is $(0.07 * 0.74) * 0.3 \text{ V} = 0.01$ W. This power can be ignored.

The total worst-case power dissipation is 1.46 W. This gives 86% efficiency for this circuit.

If we rerun the analysis for an input voltage of 12.0 V, we will see that the output diode power becomes a more significant portion of the power loss.

$$DC = (3.3 + 0.4)/12 = 0.31$$

$$P_{SWITCH} = (2.5 * 0.31 * 0.43) + (17 \text{ ns} * 2.5 * 12 * 1.25 \text{ MHz}) = 0.97 \text{ W}$$

$$P_{BOOST} = (V_O^2 * (I_{OUT}/50)/V_{IN}) = 0.05 \text{ W}$$

$$P_{DIODE} = (2.5 * (1 - 0.31) * 0.4) = 0.69 \text{ W}$$

The total worst-case power dissipation is 1.71 W. The efficiency only drops to 84% because the total loss in the switch is less due to the shorter duty cycle. Both efficiency numbers are worst-case and will be better when the IC has characteristics listed as typical in the data sheet. Also, the saturation voltage of a bipolar transistor decreases as the temperature increases, as would be expected at full output power.

Efficiency in the 85% range is adequate for systems that are powered from off-line sources, such as a desktop PC or consumer entertainment equipment. But for battery-operated equipment such as mobile phones that operate from a battery pack composed of a few cells, every extra percentage point of efficiency increases battery life. Figure 4-4 shows a buck regulator using the LT1773 synchronous controller to implement a high efficiency buck converter. The LT1773 is representative of complementary symmetry synchronous controllers available from a number of IC manufacturers.

Synchronous rectification using an NMOS transistor instead of a diode cuts losses significantly. Likewise, using a PMOS high side transistor eliminates the need for a boost supply. The top driver pulls the PMOS gate to ground to turn it on and to V_{IN} to turn it off. The bottom driver pulls the NMOS gate to V_{IN} to turn it on and to ground to turn it off. Current can flow in either direction

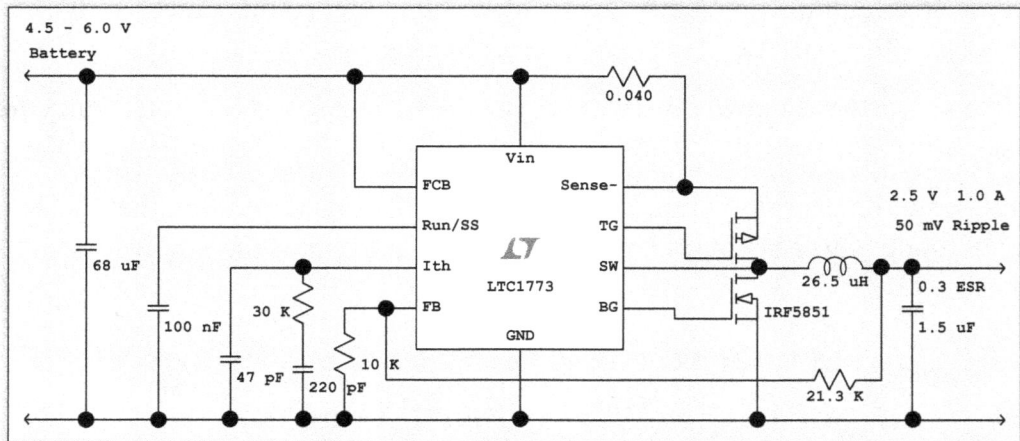

Figure 4-4: High efficiency buck converter using the LT1773 synchronous controller

through a MOSFET switch when it is turned on. The current in the NMOS switch actually flows from source to drain during normal operation. At low output current, it is possible for the inductor current to go to zero. When you use a diode, the inductor current stops as soon as the diode becomes reverse biased. With an NMOS switch, the inductor current can decrease to zero and begin to draw current from the output capacitor. The LT1773 uses the SW connection to detect when the current changes direction. When inductor current goes negative, the IC shuts off the lower switch.

The maximum input supply forces our choice of MOSFETs. The gate-source voltage will be equal to the input voltage for both MOSFETs. There are basically three classes of MOSFETs: low input, logic input, and normal input. The low input voltage MOSFETs will turn on around 1 V, but the maximum gate-source voltage is only around 8–10 V. Logic level devices usually have a maximum gate-source around 15 V and turn on around 3 V. Normal level devices have gate-source ratings around 20 V, but turn on around 4–5 V.

Synchronous rectifier controllers must ensure a minimum amount of time between turning off the top switch and turning on the bottom switch. If both transistors are on at the same time, you get a destructive short circuit from V_{IN} to ground. The inductor current must continue to flow during this dead time. The body-drain diode of the NMOS switch provides the path for the current during the dead time. This current will store charge in the diode junction until the switch turns on and then the charge will be dissipated in the switch. A small increase in efficiency is possible if the NMOS switch is paralleled with a Schottky diode. Schottky diodes do not store charge in the junction.

Figure 4-4 shows a design that is optimized for small size and low bill of materials by using a single package containing a PMOS and NMOS transistor. PMOS has roughly twice the on resistance of NMOS for the same geometry. Using individual MOSFETs for both switches will allow selecting a PMOS transistor whose resistance is roughly equal to the bottom switch. The PMOS in the IRF5851 has 0.220 Ω on resistance and the NMOS has 0.120 Ω on resistance. In our example, the power dissipation will be:

$$(1^2 * 0.220 * (2.5/6)) + (1^2 * 0.120 * (1 - (2.5/6))) = 0.092 \text{ W} + 0.07 \text{ W} = 0.16 \text{ W}.$$

The MOSFETs also draw power from the input supply as the gate is charged and discharged. Each transistor consumes current equal to the total gate charge times the frequency. The data sheet lists NMOS total gate charge as 6.0 nC at 4.5 V and PMOS gate charge of 4.5 nC at 4.5 V. We need to adjust the gate charge to account for the larger V_{GS} of 6.0 V. We have 6.0 * (6.0/4.5) = 8 nC and 5.4 * (6.0/4.5) = 7.2 nC. Total MOSFET current is 550 kHz * 15.2 nC = 8.4 mA This yields 8.4 mA * 6 V = 0.054 W used to drive the MOSFETs. The total power lost is 0.21 W, which yields 92% efficiency at maximum output. The efficiency will improve slightly as the battery discharges since there will be less power consumed driving the MOSFETs. If you use typical values for on resistance, you get 0.106 W + 0.054 W for 94% efficiency.

High efficiency controllers frequently implement burst mode for low power output situations. As the output power declines, the controller will produce a burst of pulses to charge the output and then shut off the controller while the output slowly drops to the low trip voltage where a new burst is output. This operation is very similar to how pulse frequency modulation controllers operate. Instead of a single long pulse that changes the frequency of the control, it produces one or more pulses at the fixed frequency followed by periods of no pulses. This improves EMI control because the filters only have to deal with the frequency of the oscillator.

Boost Converter Designs

Figure 4-5 shows a boost converter based on the LT1680 current mode PWM controller IC. This controller is intended for high power applications using large external NMOS switches. It includes adjustable frequency, selectable maximum duty cycle, high switch drive current, soft start, and 60 V common mode range on the current sense amplifier. The data sheet for this IC walks you through selecting all of the components needed for the design. The first selection is the operating frequency and duty cycle limit. Ours is a typical design and uses 100 kHz and 90% maximum duty cycle.

Boost converters cannot implement output short circuit protection using the control IC and the PWM circuitry. The diode provides a path from the input supply to the output independent of the switch, so the controller IC cannot turn

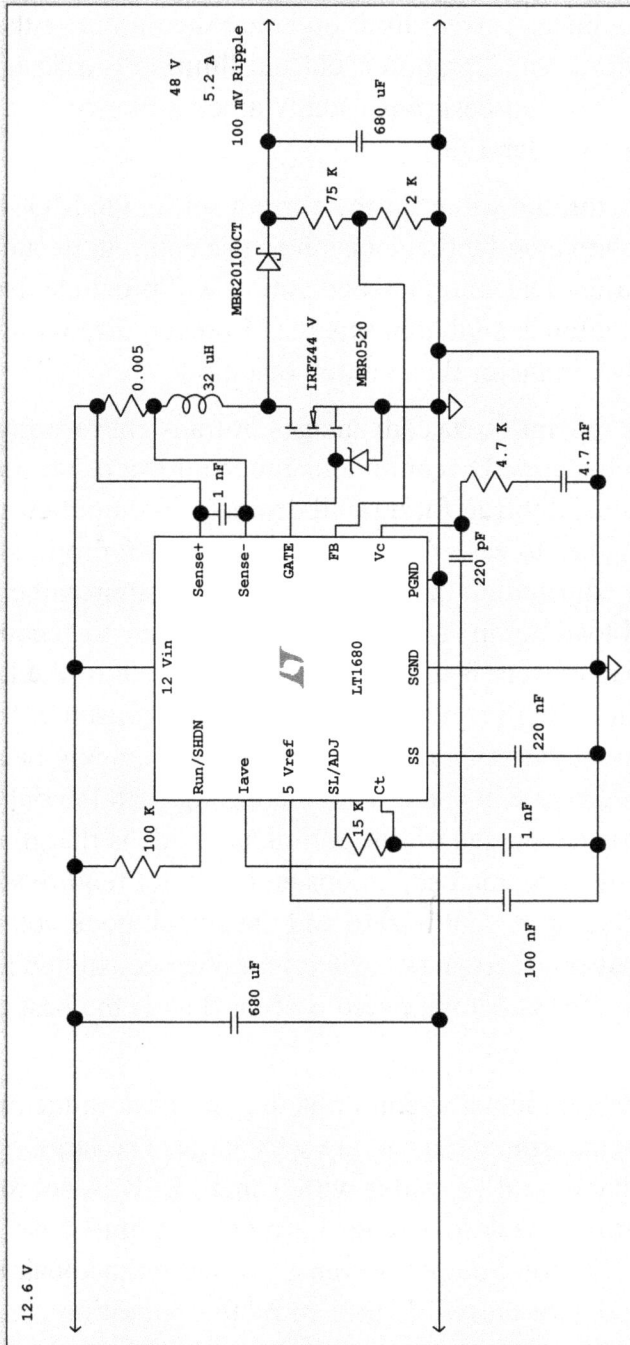

Figure 4-5: Boost converter based on the LT1680 current mode PWM controller

off current flow. The only way to implement current limiting for a boost converter is to provide a linear current limit on either the output or the input of the supply. This is a serious consideration if current limiting is a design requirement. A transformer isolated design is usually a better choice if short circuit current limiting is a consideration.

The size of the inductor, the value for the current sense, the MOSFET, and the output capacitor are affected by the decision to use continuous mode or discontinuous mode operation. Discontinuous operation will preclude using the average current limit function available in this IC. However, discontinuous mode allows using a smaller inductor than continuous mode.

Discontinuous mode operation has advantages in transient response, slope compensation, and switch losses. Discontinuous mode allows faster transient response. This is especially true for a rapid decrease in output current. Since the inductor current goes to zero for each cycle, a sudden drop in output current demand can be adjusted on the very next cycle by shortening the duty cycle. This is called load dump. The only thing necessary to accommodate a rapid decrease in load current is to draw down the current of the last pulse stored in the capacitor. There is no inductor current to consume. Likewise, a rapid increase in output current can be accommodated quickly because a large amount of the new current can come from increasing both the duty cycle and peak current. Another advantage of discontinuous mode is that the circuit will not be affected by subharmonic oscillations and will not require slope compensation. Since the inductor current is zero and the switch node voltage is zero when the switch turns on, there is no switching power consumed as the switch turns on. Turning on the switch with zero current flow is the best case for switching loss.

The down side to discontinuous operation is that the peak inductor current, peak switch current, and ripple current are very large. The large ripple current requires a larger value output capacitor with a small ESR. Also, the switch will have a very large ratio of peak to average current, so it must have a very large peak current rating. The total output power is limited by the peak inductor current, and the peak inductor current is limited by the saturation characteristic of the inductor. Once the inductor saturates, it can no longer store additional energy. The inductor current is no longer controlled by the applied voltage when saturated, so the switch current can increase very quickly. Damage to the switch is likely if the inductor saturates. The ripple voltage is a function of load current for discontinuous operation. Larger output current translates directly to larger ripple voltage.

Continuous mode operation has advantages in ripple current, peak inductor current, peak switch current, and maximum output power. Load dump for a rapid decrease in output current is problematic because all of the energy stored in the inductor must be dumped into the load. Even though the switch is turned off for multiple cycles, it is possible for the output voltage to increase quickly because of the energy stored in the inductor. The slow transient response in continuous mode makes soft start even more important. The slow response makes it very likely for large output voltage overshoot without soft start. In essence, soft start mixes a very slow transient response during startup with a faster transient response for normal operation. The lower ripple current in continuous mode allows using output capacitors with lower capacitance and higher ESR for reasonable ripple voltage. The ripple voltage in continuous mode is constant.

The switch must have a larger power rating in continuous mode service because the switch will turn on with full output voltage applied while carrying full inductor current. This is the worst case for switching losses in a switch. Continuous operation requires slope compensation for duty cycles greater than 50%. Slope compensation also requires that the inductor have a minimum value to ensure that the slope compensation stays in control. The larger inductor allows larger output power but at the expense of transient response.

Our example will use continuous mode operation since the application is a 48 V telecom application with relatively constant output power. Again, we will choose ripple current equal to 10% of the full inductor current. For the buck converter, the peak inductor current was equal to the output current plus one-half of the ripple current. This is not the case in the boost converter. We can start from the recognition that the energy stored in the inductor while the switch is closed is equal to the energy delivered to the load:

$$V_{IN} * I_{L\text{-}AVG} * DC = (V_{OUT} - V_{IN}) * I_{OUT} \tag{4-4}$$

$$DC = (V_{OUT} - V_{IN})/V_{OUT} \text{ for a boost converter.}$$

Rearranging gives average inductor current:

$$I_{L\text{-}AVG} = (V_{OUT} * I_{OUT})/V_{IN}.$$

Substituting for the maximum load condition gives:

$$I_{L\text{-}AVG} = (5.2 * 48.0)/12 = 20.8 \text{ A.}$$

The peak inductor current will be 20.8 A + one-half of the ripple current = 20.8 + 2.1 = 22.9 A.

Now we are ready to determine the size of the current sense. From the data sheet we find that:

$$R_{SENSE} = 120 \text{ mV}/I_{LIMIT}, \text{ so } R_{SENSE} = 0.12/22.9 = 0.005 \text{ } \Omega$$

Notice that R_{SENSE} is placed between the inductor and the input supply. It could also be placed between the inductor and the switch, but this would cause a problem for the current sense amplifier. Placing the resistor at the power input keeps the common mode voltage for the current sense amplifier stable and close to the supply voltage. Placing the sense resistor on the switch side of the inductor will cause the common mode voltage to change from ground to full output voltage in every cycle. The additional AC voltage caused by limited common mode rejection would disrupt proper operation of the current sense amplifier.

We can use the inductor equation to derive the boost converter equivalent of Eq. (4-1):

$$L = V_{IN} * (V_{OUT} - V_{IN})/(\Delta I * f * V_{OUT}) \qquad (4\text{-}5)$$

Substituting the values from Figure 4-5 gives:

$$L = 12.0 * (48.0 - 12.0)/(2.8 * 100 \text{ kHz} * 48.0) = 32 \text{ } \mu\text{H.}$$

The next component to pick is the switch transistor. At 100 kHz, a MOSFET is the only reasonable choice. The breakdown voltage must be higher than the output voltage. The switch needs a small amount of margin for safety. The IRFZ44V has a minimum breakdown of 60 V, which gives a 25% margin. We also have to be sure that there is adequate current capability and power dissipation.

The peak current for our design is 22.9 A, so we are well below the 39 A (100°C) continuous rating of the part. The last consideration is adequate power dissipation. The worst-case duty cycle is 90% and maximum current is 22.9 A. The device worst-case on resistance is 0.0165 Ω, so maximum power is 22.9 * 22.9 * 0.0165, or 8.7 W. The LT1680 data sheet shows that the rise and fall times will be 50 ns with the 1800 pF of the IRFZ44V. It is safe to assume the same rise and fall times from the data in the IC data sheet. The switching losses are:

$$50 \text{ ns} * I_{PK} * V_{OUT} * f = 50 \text{ ns} * 22.9 * 48 \text{ V} * 100 \text{ kHz} = 5.5 \text{ W.}$$

The total dissipation of 14.2 W is well within the capability of the switch with a proper heat sink. (See Chapter 7 for derivation of the switching loss formula.)

The peak diode current is equal to the peak inductor current, so we need a diode with a peak current rating of 23 A and breakdown voltage at least as large as the output voltage. The duty cycle of the diode current is much smaller than the switch duty cycle, so the average power is much less than the peak power. The MBR20100CT dual Schottky diode will provide more than enough margin with 100 V breakdown, 20 A current per device, and 0.9 V forward drop. Worst-case power dissipation occurs during the short time of a load dump where the full peak inductor current flows continually. This will give 0.9 V * 22.9 A, or 20.6 W. The diode duty cycle of 25% will give an average power of 5.1 W. This power will require a heat sink for the diode.

The input and output noise problems are opposite to the buck converter case. The input current for a boost converter is constant (with continuous operation) with ripple current equal to the ripple current of the inductor. This makes the filter requirements relatively easy to implement. The filter job is also easier since the waveform is a triangle wave instead of a square wave. We can approximate the RMS ripple current as 0.707 * (P-P ripple current/2). This is not exact, but we don't need an exact value. We will need margin anyway, so a minor error will just be factored into the margin.

The output current applied to the output capacitor is essentially a sawtooth wave with a peak value equal to the peak inductor current. The output capacitor ESR is very important because of the very large value of the ripple current. The RMS ripple current can be determined from:

$$I_{RMS} = I_{PK} (DC - DC^2) \qquad (4\text{-}6)$$

We can use the same process as in the buck converter design to assign one-third of the ripple voltage to the capacitor impedance and two-thirds of the ripple voltage to the ESR of the output capacitor. As in the buck converter case, we can end up needing more capacitance in order to meet the voltage ripple and capacitor dissipation requirements because of large ESR.

The boost design does not lend itself to synchronous rectification. It is possible to implement synchronous rectification, but it must use discrete components. I have found only one boost IC that implements synchronous rectification. Boost converters use a diode as the rectifier, which reduces the best possible efficiency. In applications where the duty cycle is around 50%, the diode dissipation can be much larger than the switch dissipation. The efficiency of this design is approximately 89%.

Figure 4-6 shows a battery application for a boost converter where the input is supplied by one lithium cell or multiple NiMH cells. The MAX1896 is a 6-pin IC that is designed for minimum bill of materials and extremely small size. It is a current mode PWM controller that implements all current mode functions such as slope compensation, feedback compensation, switch frequency, and current sense inside the IC. Operation at 1.4 MHz contributes to its small size because the inductor can be very small and the filter capacitors can be either ceramic or tantalum. The control circuit also implements pulse skipping to allow operation at low output current.

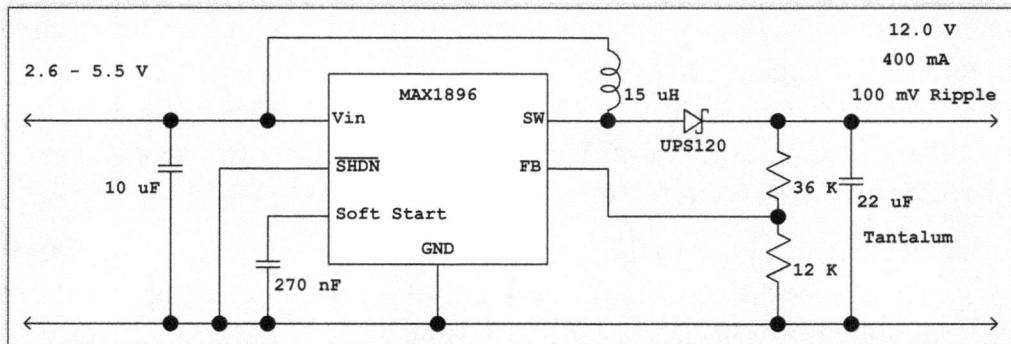

Figure 4-6: Example NiMH or lithium battery-operated boost converter

This IC takes advantage of the control of parameters that is possible in monolithic circuits. Since the FET switch on resistance is well controlled, it can be used as the current sense for the PWM circuit. The voltage at the L_X pin is directly proportional to the inductor current when the switch is on. The current sense across the 0.7 Ω on resistance sets a current trip between 550–800 mA. The current limit is a function of both on resistance and slope compensation (and, indirectly, of duty cycle).

This circuit is a little different from the previous example where we had a fairly stable input voltage. The battery voltage will change significantly during use. The voltage will decrease rather rapidly near end of charge for NiMH cells. We have to design the circuit assuming the lowest input voltage to ensure enough duty cycle to store adequate energy in the inductor. Again, we must choose the ripple current. We can choose a larger ripple current of 100 mA since the high frequency allows use of relatively small capacitors while still having small ripple voltage. We use the values from Figure 4-7 and Eq. (4-5):

$$L = 2.6 * (12.0 - 2.6)/(0.10 * 1.4 \text{ MHz} * 12.0) = 15 \text{ }\mu\text{H}.$$

The 1.4 MHz switching frequency allows very small filter capacitors that can be either tantalum or ceramic. The internal compensation circuit relies on a low frequency zero provided by a tantalum capacitor and its ESR. If a ceramic capacitor is used, the very low ESR will place the zero at a much higher frequency. The other problem with ceramic capacitors is that the equivalent inductance is significant. The inductance and small ESR complicates the loop equation. The data sheet provides the data necessary to calculate a feed forward capacitor that will externally compensate the feedback loop when using a ceramic capacitor.

This circuit has an internal soft start circuit that requires only a capacitor to set the soft start time. The circuit limits the switch current until the soft start pin reaches 1.5 V. We can use the comparator voltage and the 4 μA soft start current to calculate the capacitor value from the time needed. Our example needs 100 ms of soft start time, so we can use the definitions for capacitance, charge, and current:

$$\text{Total charge} = \text{current} * \text{time} = 4 \text{ }\mu\text{A} * 100 \text{ ms} = 400 \text{ nC}$$

$$C = Q/V = 400 \text{ nC}/1.5 \text{ V} = 266 \text{ nF}$$

So we pick a standard value of 270 nF for the soft start capacitor.

Inverting Designs

Figure 4-7 shows an inverting design using the MAX1846 inverting controller. This controller IC is intended for a full function design where maximum control of parameters is balanced with small size. This design is another in which we will create an output from a number of NiMH cells. The 3.0 V minimum input range of the controller IC precludes use in a single lithium cell application.

The first parameter to choose is the switching frequency. The controller frequency can be set from 100–500 kHz. The efficiency of the design will depend on the switch frequency because of the requirement for a P-channel FET. P-channel FETs have larger switching losses than N-channel devices because they are minority carrier devices. We need to balance the dynamic losses in the switch with the improved performance that comes with smaller components and higher frequency. The other parameter that constrains the switching frequency is the maximum duty cycle versus switching frequency. The controller has a minimum off time of 400 ns that lowers the maximum duty cycle as frequency increases. As the ratio of absolute output voltage to input voltage increases, the maximum duty cycle increases. We can rearrange Eq. (1-10) from Chapter 1 to get duty cycle as a function of input and output voltage:

$$\text{Duty Cycle} = V_{out}/(V_{out} - V_{in}).$$

Substituting our values, we see that the maximum duty cycle is:

$$\text{Duty Cycle} = (-12)/(-12 - 3.0) = 80\%.$$

The data sheet shows that the typical value for a maximum duty cycle of 80% occurs for a 500 kHz switching frequency. There will not be enough margin because the worst-case conditions for the IC will not give an 80% duty cycle. We can select a switching frequency of 400 kHz to provide the needed margin.

This example power supply is intended for an analog system that may have significant current transients, so we need to design this supply for rapid transient

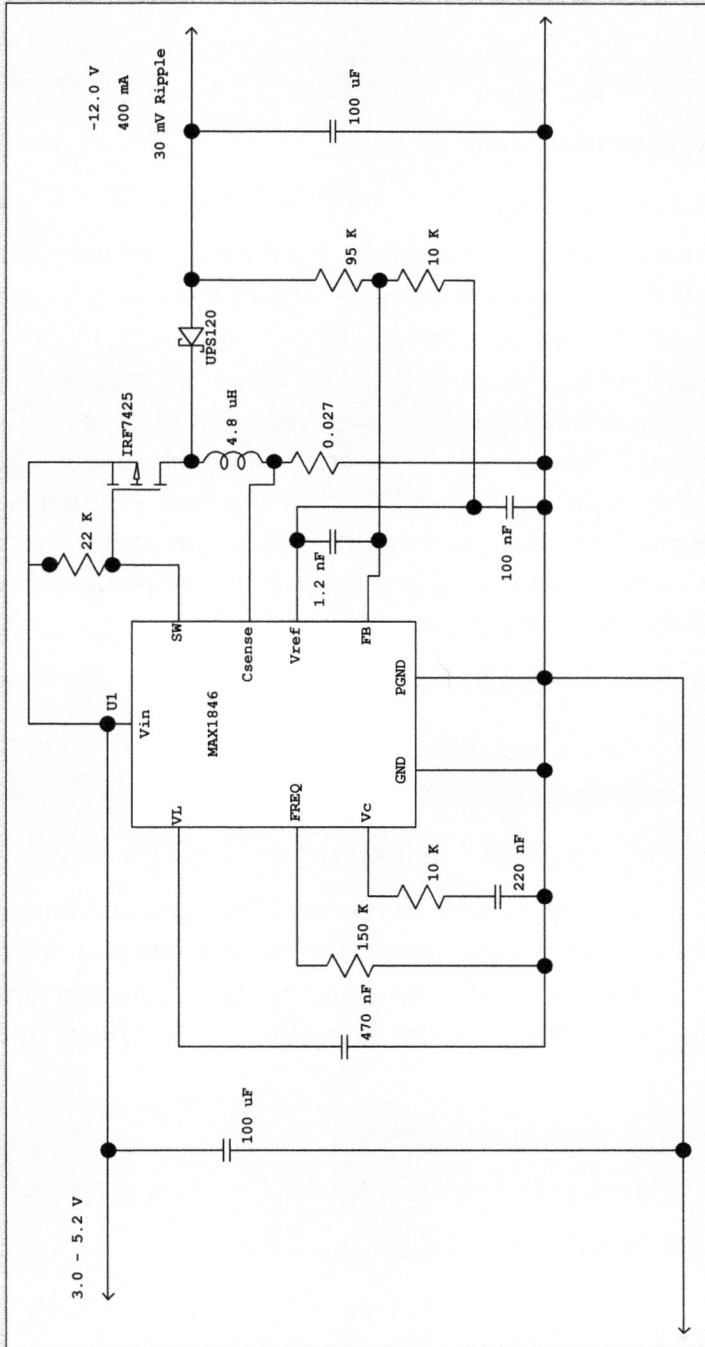

Figure 4-7: Inverting design using the MAX1846 inverting controller

response. The same relationship expressed in Eq. (4-4) applies for an inverting design.

$$V_{IN} * I_{L\text{-}AVG} * DC = V_{OUT} * I_{OUT}.$$

Substituting duty cycle and rearranging gives:

$$I_{L\text{-}AVG} = (V_{OUT} - V_{IN}) * I_{OUT}/V_{IN} = (-12 - 3.0) * -0.5/3.0 = 2.5 \text{ A}. \quad (4\text{-}7)$$

Notice that it is important to use proper signs for the output current and output voltage! We can select ripple current equal to 50% of the average inductor current at maximum load. This gives a peak inductor current of 3.13 A. Setting such a high ripple current at maximum load and minimum input voltage is likely to cause discontinuous operation at very light loads and at maximum input voltage. The ripple voltage will be constant while the supply operates in continuous mode. The ripple voltage will be even less once the supply changes to discontinuous operation at low output current. It is important to verify loop stability in the lab for both continuous and discontinuous operation because the loop gain will be different depending on the mode.

We again use the inductor equation and rearrange for an inverting converter:

$$L = (V_{IN} * V_{OUT})/((\Delta I * f) * (V_{OUT} - V_{IN})). \quad (4\text{-}8)$$

Substituting the values from Figure 4-8 gives:

$$L = (3.0 * (-12.0))/((1.25 * 400 \text{ kHz}) * (-12 - 3.0)) = 4.8 \text{ μH}.$$

Figure 4-8 gives the operating parameters for both the input and output. We used the minimum input voltage and maximum output current to determine the size of the inductor, assuming continuous mode operation. We can use the maximum input voltage and minimum output current to see the effects of discontinuous operation.

The duty cycle for fully charged cells will be:

$$\text{Duty Cycle} = (-12)/(-12 - 4.2) = 0.74.$$

We can rearrange Eq. (4-8) to solve for ΔI:

$$\Delta I = (V_{IN} * V_{OUT})/((L * f) * (V_{OUT} - V_{IN})) = (4.2 * (-12))/$$
$$((4.8 \text{ μH} * 400 \text{ kHz}) * (-12 - 4.2)) = 1.62 \text{ A}.$$

$I_{L\text{-AVG}}$ is equal to $\Delta I/2$ at the point where the mode changes from continuous to discontinuous. We can rearrange Eq. (4-7) to solve for I_{OUT}:

$$I_{OUT} = (I_{L\text{-AVG}} * V_{IN})/(V_{OUT} - V_{IN}) = (0.81 * 4.2)/(-12 - 4.2) = 210 \text{ mA}.$$

This result indicates that our design will be in discontinuous mode when the cells are fully charged and the load current is less than 210 mA.

An inverting supply implemented with a current mode IC will give inherent short circuit current limiting because the switch disconnects the inductor from the input supply. The output short circuit current will be limited to the peak inductor current.

The current sense resistor is determined from the peak inductor current at maximum output using the formula from the data sheet:

$$R_{CS} = 0.085 \text{ V}/I_L = 0.085/3.13 \text{ A} = 0.027 \ \Omega.$$

The switch gate voltage is equal to the input voltage. This means we must start our search for a device that will fully turn on at 3.0 V. The drain-source voltage will be equal to the input voltage plus the output voltage, so the breakdown voltage must be greater than 16.2 V. The last parameter to determine is peak drain current. Our supply has a peak inductor current of 3.13 A. The IRF7425 is a reasonable device that meets these requirements.

Both the input capacitor current and the output capacitor current are discontinuous for an inverting design. The waveform on the input and output are sawtooth waves with peak amplitude equal to the peak inductor current. Both ESR and the ripple current rating are important considerations for the input and output capacitors. The input filter considerations of Q and negative input impedance are the same as we saw with the buck converter. The RMS input capacitor current is shown below:

$$I_{RMS} = I_{OUT} (DC/(1 - DC))^{1/2}$$

Step Up/Step Down (Buck/Boost) Designs

Figure 4-8 shows an implementation of a step up/step down design based on the MAX641 buck/boost converter.

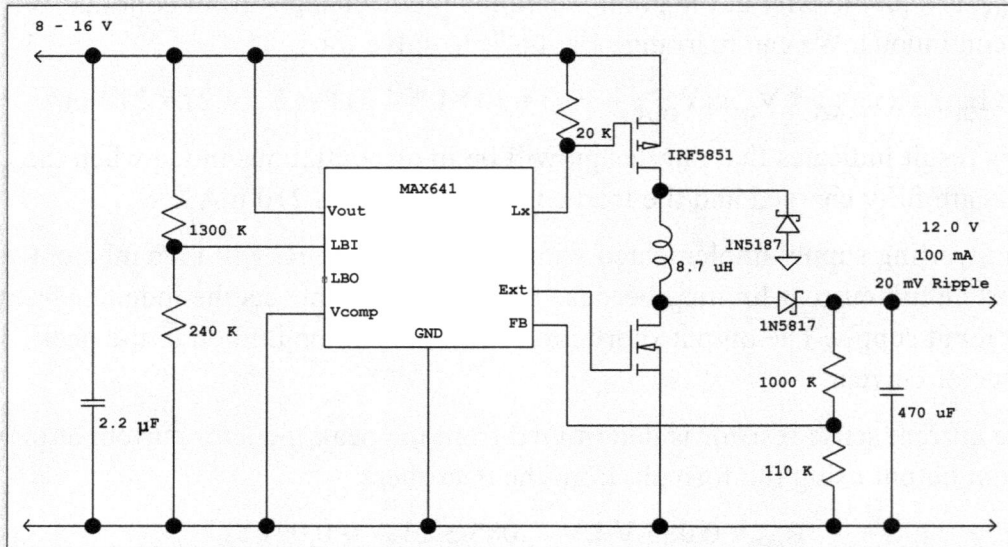

Figure 4-8: Step up/step down design based on the MAX641 buck/boost converter

The MAX641 is intended as a step-up regulating converter for fixed output use. It lends itself to the buck/boost design because it has complementary drive pins. The L_x pin is driven from an internal MOSFET and the Ext pin is intended to drive an external MOSFET switch in higher power designs. The V_{OUT} pin is actually used to supply power to the IC internal circuits. In normal boost mode, the inductor will supply current to bootstrap the system. In this design, we need to use the V_{OUT} pin tied directly to V_{IN} to supply IC current.

The design method is identical to the inverting design. The formula for the duty cycle for continuous mode is:

$$\text{Duty Cycle} = V_{OUT}/(V_{IN} + V_{OUT}).$$

The maximum inductor current will occur when the input voltage is below the output voltage and the system is acting as a boost converter. The same relationship expressed in Eq. (4-4) applies for this design.

$$V_{IN} * I_{L\text{-}AVG} * DC = V_{OUT} * I_{OUT}.$$

Substituting duty cycle and rearranging gives:

$$I_{\text{L-AVG}} = (V_{\text{OUT}} + V_{\text{IN}}) * I_{\text{OUT}}/V_{\text{IN}} = (4.0 + 6.0) * 1.0/4.0 = 2.5 \text{ A}.$$

We can select ripple current equal to 20% of the average inductor current at maximum load. This gives a peak inductor current of 2.75 A. We will need to use the minimum input voltage to select an inductor small enough to allow adequate current flow.

We again use the inductor equation and rearrange for this converter:

$$L = (V_{\text{IN}} * V_{\text{OUT}})/((\Delta I * f) * (V_{\text{OUT}} + V_{\text{IN}})).$$

Substituting the values from Figure 4-9 gives:

$$L = (4.0 * 6.0)/((0.5 * 550 \text{ kHz}) * (4.0 + 6.0)) = 8.7 \text{ }\mu\text{H}.$$

The selection of the MOSFETs follows the criteria we have used before. First, we select parts based on the gate voltage and then based on the drain voltage. Finally, we ensure that the current rating is adequate.

This design is quite costly because of the extra switch and the extra diode required. These additional components add cost to the bill of materials and they reduce the efficiency. Another design that trades the cost of the extra switch and diode for the cost of an extra inductor and capacitor is the single-ended primary inductance converter (SEPIC) design. A representative SEPIC design is shown in Figure 4-9.

There is very little design information available on SEPIC converters. In my search of IC vendor websites, I found three barely adequate descriptions of the design of a SEPIC converter. These were DN48 from TI (Unitrode), AN1051 from Maxim, and an article by National Semiconductor in *EDN Magazine*, October 17, 2002.

The best way to describe the operation of the SEPIC converter (and the Cuk and Zeta converters, which are variants) is to begin thinking of the circuit as similar to an RC voltage coupled amplifier stage. In an RC amplifier, the load resistor allows the active device (a switch, in our case) to produce a varying voltage by changing the amount of current drawn through the resistor. This AC voltage is coupled to the load circuit by the capacitor, which is a short circuit for the AC. The capacitor blocks the DC applied to the amplifier from the load. At RF, the resistor can be replaced with a choke so that the amplifier dissipates

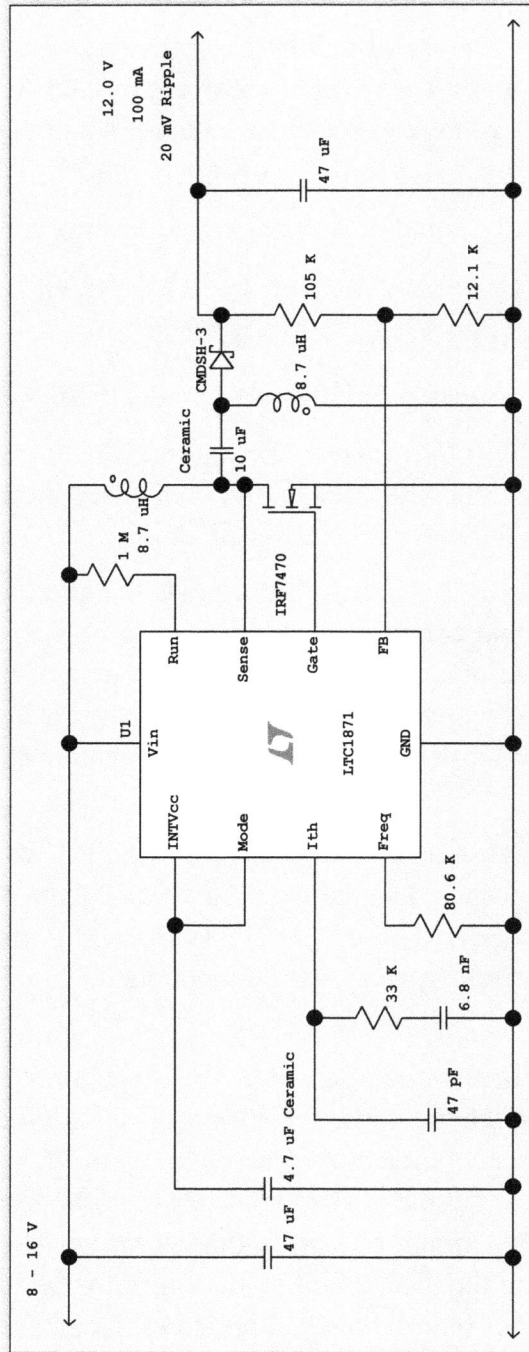

Figure 4-9: Representative single-ended primary inductance converter (SEPIC) design

less power. The square waveform is transferred to the load circuit where the diode, inductor, and filter capacitor convert the AC into the DC output voltage, just as in a buck converter.

The two inductors in a SEPIC circuit have the same voltage and the same current as long as the inductor values are identical. SEPIC circuits are usually designed with equal value inductors to simplify the design, but equal value inductors are not required. If the inductor values are equal, both inductors can be wound on the same core. DN 48 gives a reasonable design sequence. The inductor and duty cycle equations for a SEPIC circuit are the same as for the buck/boost design in Figure 4-8.

First, we select the switching frequency. Next, we calculate the highest inductor current. The maximum inductor current will occur when the input voltage is below the output voltage and the system is acting as a boost converter.

$$I_{L\text{-}AVG} = (V_{OUT} + V_{IN}) * I_{OUT}/V_{IN} = (4.0 + 6.0) * 1.0/4.0 = 2.5 \text{ A.}$$

We can select ripple current equal to 20% of the average inductor current at maximum load. This gives a peak inductor current of 2.75 A. We will need to use the minimum input voltage to select an inductor small enough to allow adequate current flow.

We again use the inductor equation and rearrange for this converter:

$$L = (V_{IN} * V_{OUT})/((\Delta I * f) * (V_{OUT} + V_{IN})).$$

Substituting the values from Figure 4-9 gives:

$$L = (4.0 * 6.0)/((0.5 * 550 \text{ kHz}) * (4.0 + 6.0)) = 8.7 \text{ } \mu\text{H.}$$

This is the value for both inductors.

The next step is to determine the RMS ripple current in the coupling capacitor. DN48 gives the following equation:

$$I_{C \text{ RMS}} = (I_{OUT}(max)^2 * DC \text{ (max)} * I_{IN}(max)^2 * (1 - DC(max)))^{1/2}$$

$$= (1 * (4/(4+6)) * 2.5^2 * (1 - (4/(4+6))))^{1/2}$$

$$= 1.22 \text{ A}$$

We must select a coupling capacitor that has the power handling ability for this ripple current.

We pick the output capacitor to give the desired output ripple, again allocating two-thirds of the ripple voltage to ESR and one-third to X_C. The output ripple current is given by:

$$I_{RMS} = I_{OUT} (DC/(1 - DC))^{1/2}$$

We use the equations shown in the buck converter section to calculate the capacitor value and determine the required ESR.

The gate voltage of the MOSFET is set by the internal voltage regulator to 5.2 V. This will require a logic level MOSFET. The drain voltage is equal to $V_{IN} + V_{OUT}$. The peak MOSFET current is $I_{IN} + I_{OUT}$. The peak diode current and peak reverse voltage are equal to the MOSFET voltage and MOSFET current.

Charge Pump Designs

The design sequence below should serve as a starting point for those new to charge pump design:

1. Choose an IC based on the output power, physical size, input voltage, etc. The ratio of input voltage to output voltage will determine if a step-up, step-down, or inverting converter is selected. You also select based on whether output voltage regulation is required.

2. Choose the switch frequency (if adjustable) and flying capacitor value.

3. Choose the output ripple voltage. Choose the output capacitor based on the output ripple voltage.

Figure 4-10 shows a step up charge pump with output regulation. The LTC3200-5 will produce a regulated 5.0 V over the input range of a single lithium cell. This circuit is typical of charge pump ICs where the number of external components is very small. This IC requires just three capacitors to convert an unregulated low voltage into regulated 5.0 V. The IC provides an internal voltage divider for the feedback circuit to set the output voltage to 5.0 V.

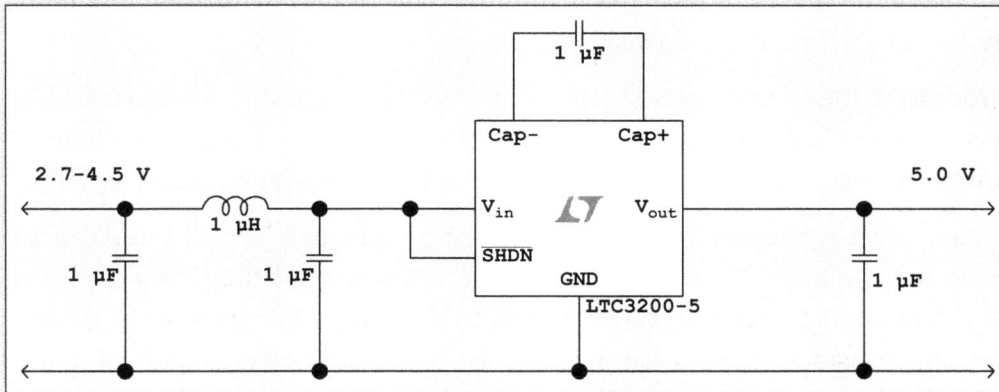

Figure 4-10: Step up charge pump with output regulation

Small ceramic capacitors are indicated for all three capacitors because of the 2-MHz switch frequency.

As mentioned in Chapter 1, the charge pump has an equivalent resistance of:

$$R_{EQ} = 1/(f * C_{FLYING})$$

as long as the equivalent resistance is much larger than the internal switch resistance. This equivalent resistance is one source of power dissipation. It is possible to reduce the dissipation by using a larger value flying capacitor until the ESR dominates.

The data sheet warns against using tantalum or aluminum electrolytic flying capacitors because the flying capacitor can have negative voltage during the startup of the supply. These capacitors would not be a good choice for this supply in any case, because they will have significant ESR at the 2-MHz switch frequency.

The efficiency of the supply depends on the ratio of output and input voltage. The switching nature of the charge pump would yield nearly 100% efficiency for an output voltage twice the input voltage and very light load. As the load increases, the losses in the ESR of the capacitor and the internal resistances increase. The losses due to the equivalent switching resistance also increase as load current increases. These losses limit the output current at very low input voltages, as shown on the data sheet.

In order to regulate to a voltage less than twice the input, the IC must dissipate power for input voltages from 2.7–5.0 V. The IC operates much like a linear regulator in this case. At input voltage of 5.0 V, the efficiency drops to 50%.

The data sheet suggests that 1-μF capacitors are adequate for all three capacitors. Maximum output current is dependent on the size of the flying capacitor until the internal switch resistance begins to dominate. The data sheet gives equivalent resistance versus temperature for two input voltages and using a 1-μF flying capacitor. The switching equivalent resistance is $1/(f * C)$, which is 0.5 Ω.

Next, we choose the output capacitor based on the required output ripple. We will use a ceramic capacitor that has essentially zero ESR, so the capacitive reactance is the dominant source of ripple.

$$V_{p\text{-}p} = I_{OUT}/(2 * \pi * f * C) = 40 \text{ mA}/(6.28 * 2 \text{ MHz} * 1 \text{ μF}) = 3 \text{ mV}$$

We use the value of I_{OUT} because the duty cycle is 50% and the current supplied to the output is essentially a square wave.

The value of the input capacitor has less effect on the ripple because the input current is essentially equal while the flying capacitor is charging and while current is transferred to the output. There is a very short period of time where the nonoverlapping clocks that drive the switches are all off. This time is approximately 25 ns for the LTC3200-5. The RMS value of such a short pulse is very small. However, the rise and fall times are still rather fast, so the capacitor must be very close to the IC to prevent the inductance of the input trace from creating a resonant circuit that will ring. Figure 4-10 shows an inductor and additional capacitor forming a pi filter to provide additional filtering of noise reflected to the input supply.

Figure 4-11 shows an inverting charge pump with output regulation. Again, this IC provides a regulated output voltage with a small component count (four capacitors and two resistors). This IC works by charging the two flying capacitors in parallel across the input supply during the charge phase. During the

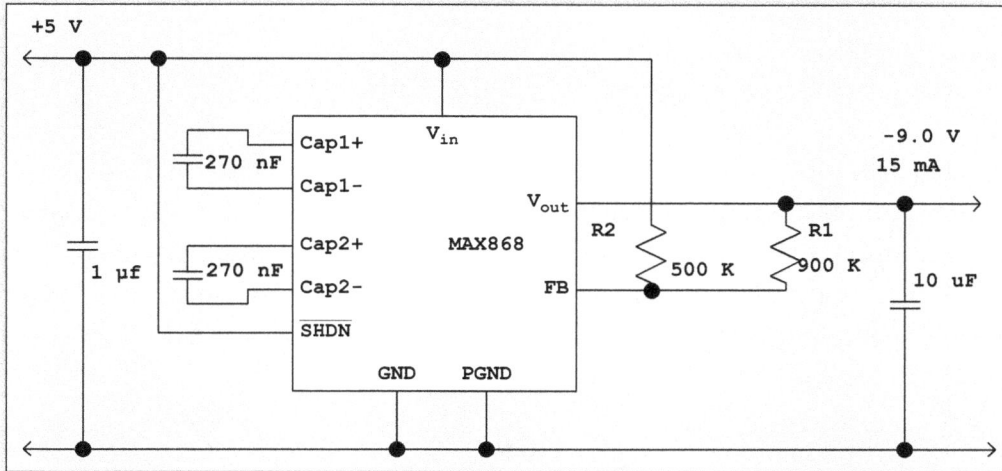

Figure 4-11: Voltage inverting charge pump with output regulation

discharge phase, the switches are reconfigured to stack the two flying capacitors in series to produce a negative voltage equal to twice V_{IN}.

Maxim describes the regulation mechanism as PFM control, but the control mechanism is actually pulse dropping. The clock runs at a constant 450 kHz and the control circuit drops pulses as required to keep the output in control.

You will notice that one feedback resistor is tied to the input voltage. This is required because the feedback pin must be above ground. The design in Figure 4-11 requires the input voltage to be regulated because the feedback circuit uses the input voltage as the reference. An alternative is to provide a second input voltage that is a reference for the regulator. The data sheet suggests setting R2 to a value between 100–500 K to limit the current draw from the divider. Then use the formula from the data sheet for R1:

$$R1 = R2 * (|V_{OUT}|/V_{REF}).$$

The data sheet provides equations that allow us to calculate the required values for C1, C2, and C_{OUT}.

For a given value for C1 and C2, we can verify that the maximum output current available meets the design goal. Maxim chooses to give this equation rather than a formula to calculate the capacitors:

$$I_{OUT(MAX)} = \frac{(2 \times V_{IN}) - |V_{OUT}|}{\dfrac{4}{f_{max} \times (C1 + C2)} + R_{OUT} \times \dfrac{10\ V}{V_{IN} + |V_{OUT}|}} \qquad (4\text{-}9)$$

The data sheet gives 450 kHz as the maximum frequency and 70 Ω as the equivalent output resistance. Equation (4-10) is the result of Eq. (4-9), based on setting C1 and C2 to infinity. This will give the absolute maximum current based only on the voltages and the IC characteristics. We substitute 5.0 V input and −9.0 V output to determine if we will be able to obtain 15 mA current out.

$$I_{OUT(MAX)} = \frac{(2 \times 5.0) - |-9.0|}{70 \times \dfrac{10\ V}{5.0 + |-9.0|}} \qquad (4\text{-}10)$$

This indicates we should be able to use this device for this amount of current. The denominator of Eq. (4-10) is 50, so we can work backward to obtain reasonable values for C1 and C2. For 15 mA of output current, the denominator in Eq. (4-9) is:

$$66.6 = 1/0.015\ mA$$

The capacitance term in the denominator of Eq. (4-9) can now be evaluated using:

$$66.6 = \frac{4}{450\ kHz \times (C1 + C2)} + 50$$

Solving yields:

$$C1 = C2 = 0.27\ \mu F.$$

The data sheet also gives an equation in terms of X_C, R_{OUT}, and ESR:

$$V_{RIPPLE(P\text{-}P)} = ((2 \times V_{IN}) - |V_{OUT}|) \times \left(\frac{1}{1 + \dfrac{4 \times C_{OUT}}{C1 + C2}} + \frac{ESR}{R_{OUT}} \right)$$

Since R_{OUT} is 70 Ω, ESR of a ceramic capacitor will not contribute to that term. 10 μF is a reasonable start for C_{OUT}. This yields ripple voltage of:

$$V_{RIPPLE(P-P)} = ((2 \times 5.0) - |\ 9.0\ |) \times \left(\cfrac{1}{1 + \cfrac{4 \times 10\ \mu F}{0.27\ \mu F + 0.27\ \mu F}} \right) = 13\ mV.$$

Tantalum capacitors will have ESR on the order of 0.5–3 Ω in this capacitance range and voltage range (depending on manufacturer and technology), so the ripple voltage will be significantly larger with tantalum capacitors.

The input capacitor ESR is much more important for an inverting supply because the IC draws current only while charging the flying capacitors. The peak input current is double the output current. The input ripple is even more important if V_{IN} is used as the reference. Once again, a large value ceramic capacitor with low value ESR is appropriate.

Layout Considerations

The basic white protoboard you used in your beginner EE classes will work for a *small* power supply up to perhaps 20 kHz switching frequency. Not many useful power supplies run at such a low frequency any more. A modern switching regulator will run from 100 kHz up to several MHz. The harmonics of the switching waveform extend up to the VHF frequency range. Failure to use a PC board that uses good high frequency layout will guarantee disappointing results (and, likely, lots of smoke).

There are two issues that we have to consider. The first is to design the layout of the power supply circuit so it does not interfere with its own operation. The second is to consider how the voltages and potentially huge current densities can interfere with the rest of the system if the power supply is placed too close to sensitive circuits.

Pentium CPUs can draw 40 A. Even 10 mΩ will produce a voltage drop of 0.4 V. In such a power supply, it is very important to keep low level signals isolated from the high current paths of the rectifiers and switches. It is easy to overlook the magnetic consequences of such currents. Each loop where this current flows is a single turn inductor that we tend to ignore. Our example would create as much as 10 A-Turns of AC magnetic field that can easily couple into adjacent traces and loops in the power supply and other close circuits. Pentium applications are rather extreme, but they illustrate how easy it is to have otherwise inconsequential layout choices become important in switching supplies.

Figure 4-12 shows a representative PCB layout and schematic from the LT1871 data sheet. This gives a good example of the considerations in layout of a circuit. The figure does not show the bottom of the PCB. The layout needs a large continuous ground plane on the bottom of the board that extends from the right side of the board to the area of the via at the IC ground pin. The ground plane should narrow at this point and then expand to connect to the vias for the timing and measurement circuits. This is indicated in the schematic by the narrower ground connection between the GND pin and the components on the left in the schematic.

The first consideration of layout is to realize that the ground current of the input supply flows directly to the output circuit. Notice that the schematic has been drawn to roughly show how components will be physically placed on the PC board. All of the switch components, as well as C_{IN} and C_{OUT}, are placed near each other and away from the signal ground connection of the LT1871. The ground connection of the IC is part of the signal measuring circuit, so any voltage changes due to switching currents flowing from the input capacitor to the output capacitor can change the voltage applied to the sense circuits inside the IC. The ground current out of the IC can also be fairly large during the times that the MOSFET is switching. The peak gate current can be on the order of hundreds of mA at turn-on and turn-off of the switch. This indicates that a fairly large trace is needed between the GND pin of the IC and the common connection between C_{IN} and C_{OUT}. Notice that the top ground area is large and the IC ground pin is at one corner of the ground area to limit the voltage change due to AC current flowing in the ground area from C_{IN} and C_{OUT}. The majority of the DC current flow in this design flows on the ground plane on the bottom of the board (not shown in Figure 4-12). The figure shows the connections for V_{IN}, V_{OUT}, and GND. The input and output ground connections should

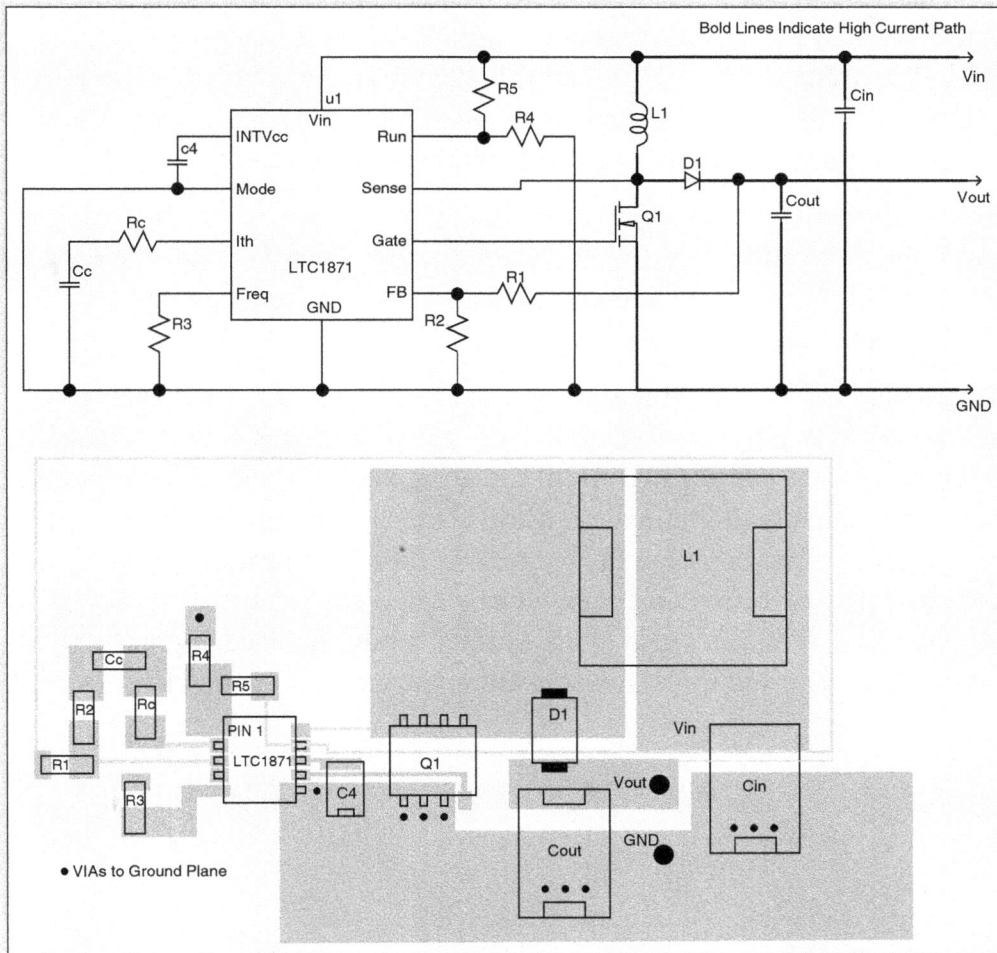

Figure 4-12: Representative PCB layout and schematic using the LT1871

be made between C_{IN} and C_{OUT} so that current flow is concentrated near the vias for the switching components.

The connections to the feedback resistors and the current sense input should be routed as far away as possible from the lines that drive the switch gate and the lines that connect the switches and the inductor. Again, there are large AC currents flowing in these traces and even small closed circuits nearby will be one-turn inductors that can produce sizeable voltages and disturb linear parts of the circuit. There are two major magnetic loops. The first is composed of L1, C_{IN}, and Q1. The second is C_{OUT}, D1, and Q1. We minimize magnetic pickup by the measuring circuits by keeping the traces small and as close together as possible. This minimizes the loop area and the induced voltage.

These same considerations apply to charge pump circuits where switching currents can be quite large. You will want to keep the common connection for the IC, C_{IN}, and C_{OUT} close and keep loops away from the feedback input if the converter is regulated.

It is important to use wide traces as much as possible at the frequencies of modern switching supplies. Even one-half inch of a narrow trace can have inductance of many tens of nH. All of the design rules in this chapter presume reasonable circuits with minimal parasitic elements. If you inadvertently design in parasitic inductances on the PC board, it is possible to create unintended additional voltage stresses on components when elements switch on or off. Where possible, it makes sense to use surface mount rather than through-hole components to help minimize the parasitic inductances in component leads.

CHAPTER 5

Transformer Isolated Circuits

- Feedback Mechanisms
- Flyback Circuits
- Practical Flyback Circuit Design
- Off-Line Flyback Example
- Non-Isolated Flyback Example
- Forward Converter Circuits
- Practical Forward Converter Design
- Off-Line Forward Converter Example
- Non-Isolated Forward Converter Example
- Push-Pull Circuits
- Practical Push-Pull Circuit Design
- Half Bridge Circuits
- Practical Half Bridge Circuit Design
- Full Bridge Circuits

Transformer Isolated Circuits

In this chapter, we will look at detailed designs of transformer isolated converters. The primary application is off-line power supplies, but these designs are also useful in applications where safety isolation is required or where the input voltage can vary above and below the output voltage.

All of the designs shown here use current mode PWM control, just like the designs in Chapter 4, because of its inherent advantages in loop stability and current control.

Feedback Mechanisms

The following section applies to transformer circuit applications where the transformer is used for isolation, such as in off-line supplies. The output can be connected directly to the control IC in applications where isolation is not required.

Most transformer circuits use the magnetic circuit of the transformer to provide electrical isolation of the secondary circuit from the primary circuit. Putting the control IC on the input side of the supply requires that feedback of the output voltage to the control IC has to cross an isolation barrier. If the IC is powered from an isolated supply, then the switch control must cross the isolation barrier.

Using an optocoupler is the "easiest" way to transfer output voltage information across the isolation barrier to a control IC on the primary side. Optocouplers, in general, provide isolation of 2500 V or more between the LED and the photo transistor. There are several characteristics of optocouplers that make them less than ideal in this application. However, they are still a reasonable choice for this application because they are small and inexpensive when compared to transformers. The first problem is the large variation in transfer function from unit to unit. This change in current transfer ratio causes a large variation in the loop equation from unit to unit. The control loop must be designed conservatively in order to account for the worst-case optocoupler. This results in a nominal system being damped more heavily than necessary.

Another problem is the low corner frequency of the transfer function. Optocoupler phototransistors are built with a rather large base region to improve the conversion of light to current. The large base region creates a larger input capacitance and reverse transfer capacitance than in regular transistors. Although it is only a few picofarads, the Miller effect will amplify the capacitance to a much larger value. The phototransistor is used in a manner identical to an RC-coupled amplifier. The Miller capacitance creates a pole at a fairly low frequency. Just as in an RC amplifier, the frequency response can be improved by using a low collector resistance. This lowers the voltage gain of the optocoupler. Agilent, Clairex, and other manufacturers produce optocouplers with better frequency response, but they are significantly more expensive than ordinary devices such as the 4N27.

The usual method of compensating for the low optocoupler gain and the capacitance of the optocoupler is to use an amplifier and voltage reference on the isolated side of the supply. National Application Note AN-1095 gives a detailed design method with rigorous analysis of the control loop for an optocoupler isolated system. Figure 5-1 shows a common drive circuit using the TL431 shunt regulator. Resistors R1 and R2 divide the output voltage down to 2.5 V for the control pin of the TL431. There are two optional compensation circuits in Figure 5-1. These can be used to add a pole or a zero to the loop response. The TL431 and its variations provide the voltage reference, comparator, and power amplifier in one convenient package. See Chapter 11 for an example of an optocoupler and op-amp used for isolated feedback. The feedback pin of the control IC is connected to the common of the input to force the IC to the largest duty cycle possible. The V_{COMP} pin is an open collector style output with a current source on most modern ICs. The resistor and capacitor to ground add more compensation and the optocoupler transistor reduces the error amplifier output to reduce the duty cycle.

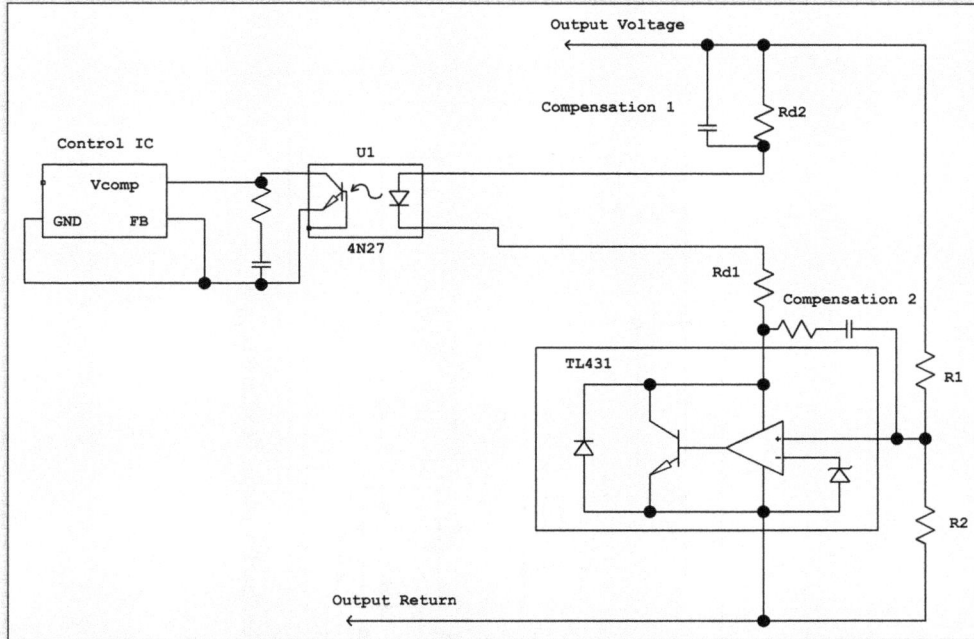

Figure 5-1: Representative optocoupler feedback with compensation circuits

Another method of feedback isolation is to use a small power line transformer to generate an isolated auxiliary supply for the IC. The IC then drives a pulse transformer to supply isolated drive for the switches. Even for fairly high output systems, the power required for switch drive and the control IC is only a few watts. The auxiliary transformer does not need to be especially large, but it must be able to switch from 110 VAC to 240 VAC. The main drawback of this method is that the transformer adds size to the supply. It is entirely possible that this auxiliary transformer could be larger than the switching transformer at the 100 W level! This method works for systems where 110 V or 240 V operation is selected manually. It is less desirable for universal input supplies because the transformer has to be able to handle 240 VAC/50 Hz nominal input but still generate enough power at 90 VAC. The only practical way to power the control IC in such a universal supply is to provide some form of linear regulation like a zener diode or three-terminal regulator. Figure 5-2 shows a representative transformer drive with an auxiliary supply. T1 is a small iron core power

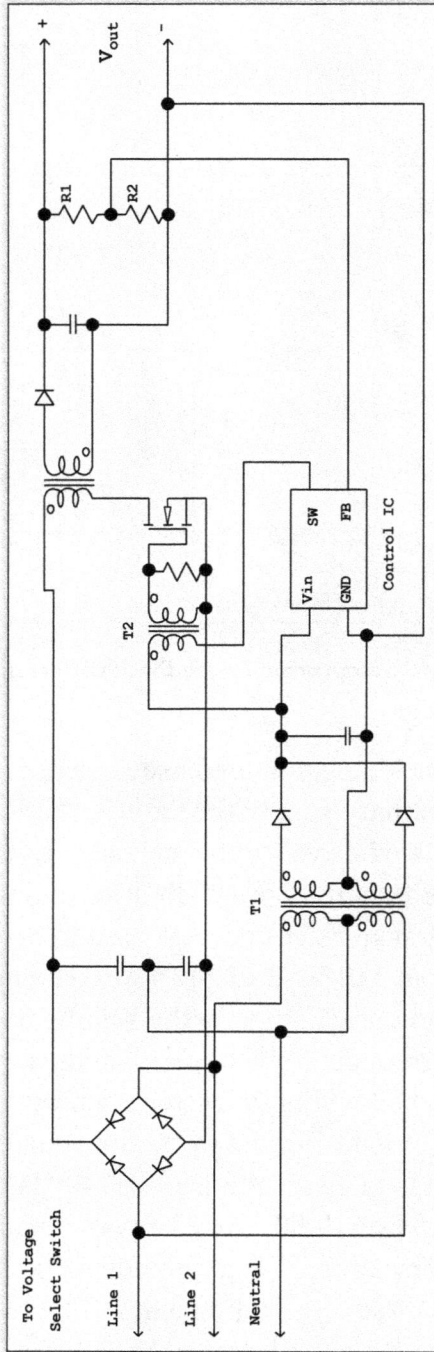

Figure 5-2: Representative isolation and feedback with an auxiliary supply

transformer and T2 is a pulse transformer for driving the MOSFET switch. Both T1 and T2 must meet safety agency isolation specifications.

TI produces ICs for use on the secondary side that use amplitude modulation of an AC signal to transfer the control signal across the isolation barrier. The UC1901 varies the amplitude of an RF carrier frequency, which is fed to a transformer and then rectified on the primary side to supply the feedback voltage. Figure 5-3 shows an application of this IC. The RF oscillator can operate up to 5 MHz. The high frequency allows the time constant of the rectifier filter (R4, C4) to be quite short so that there is minimal phase shift through the RF to DC part of the circuit. This IC also includes the error amplifier and other support circuitry. The error amplifier has a compensation pin that can be used to add poles or zeros to the loop response. TI describes this IC and applications in Application Note AN-94. The feedback transformer must meet safety agency isolation requirements similar to those imposed on the main power transformer. See Chapter 9 for more information on safety agency requirements.

An alternative to the UC1901 is to use a regular PWM control IC operating at a high frequency to drive a pulse transformer and pulse averaging circuit. Figure 5-4 shows an example. The high frequency allows the low pass filter (R1, R4, C4) to use a small capacitor so that the pulse averaging does not add significant time delay to the feedback circuit. A time delay corresponds to adding a pole to the loop response. We use C2 on the primary of the pulse transformer to avoid problems with current in the magnetizing inductance. Capacitors C3 and D3 form a DC restoration circuit. Without the DC restoration circuit, the DC level of the pulses will vary with duty cycle because the volt-microseconds of each portion of the pulse waveform will be equal. We will look at the equal volt-time characteristic transformers in detail in Chapter 7 when we look at methods of driving the switch.

Figure 5-5 shows the operation of a regular detector circuit and operation of a DC restoration circuit for three different duty cycles. In the regular detector, the output will be the height of the waveform above zero (the dark line). The area of the two shaded areas is equal, showing equal volt-seconds for the positive and negative portions of the AC waveform. The DC restored circuit below

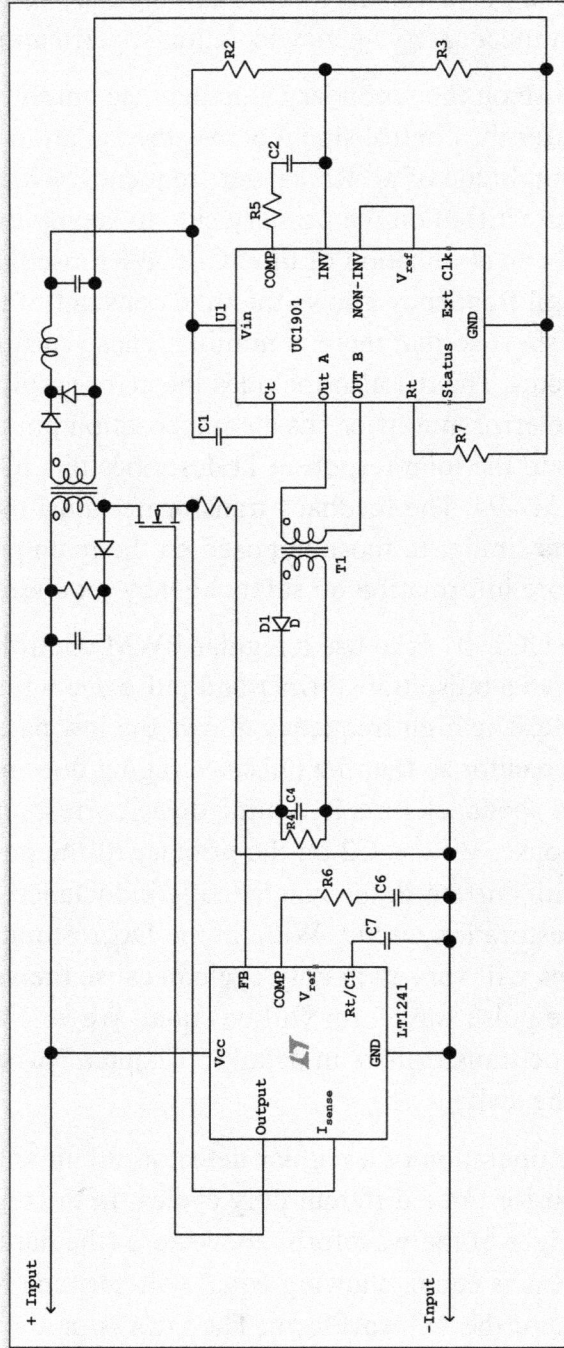

Figure 5-3: Isolated feedback using Texas Instrument's UC1901

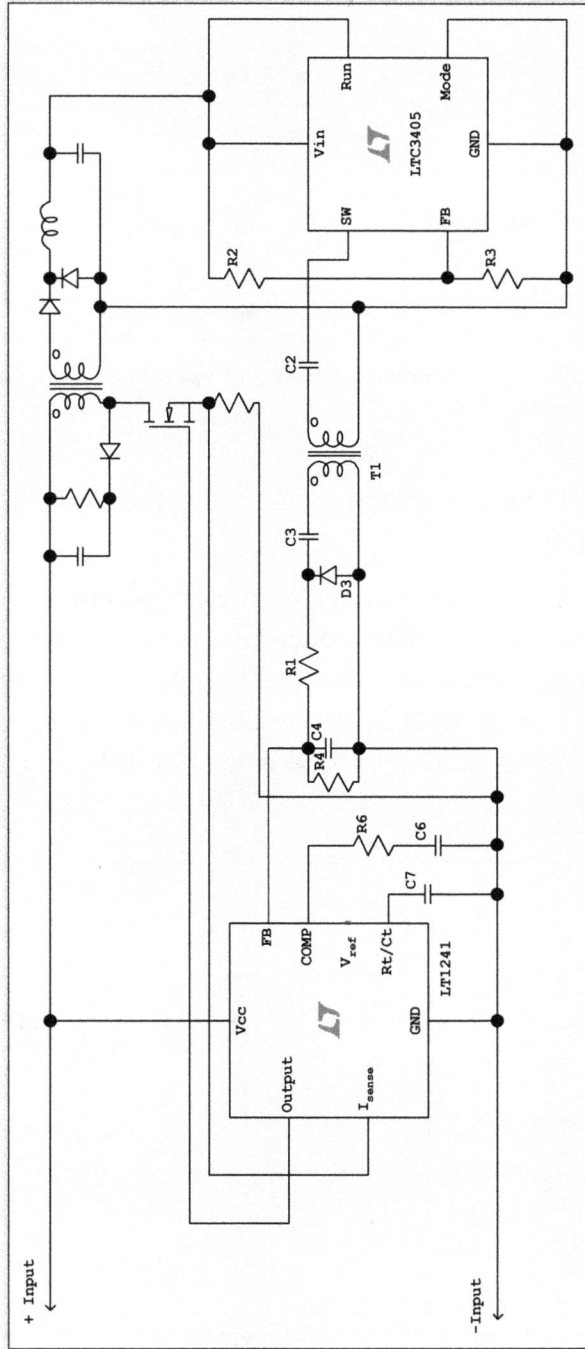

Figure 5-4: Using a standard PWM control IC for isolated feedback

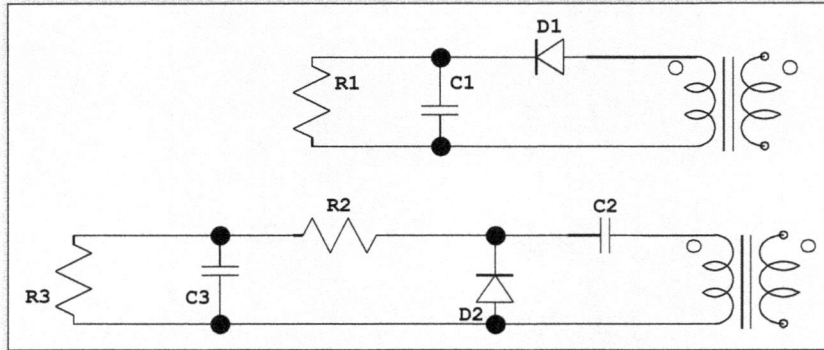

Figure 5-5: Operation of a regular detector circuit and operation of a DC restoration circuit for three different duty cycles

shows that the output is the peak amplitude minus the forward voltage of the diode for all three duty cycles.

Flyback converters maintain the voltage on the output circuits in proportion to the turns ratios of the inductor. The inductor winding charges each output capacitor to the voltage across the winding. This property allows use of a secondary circuit to provide both IC power and output voltage measurement. Figure 5-6 shows a representative flyback converter. D1 and C2 provide an

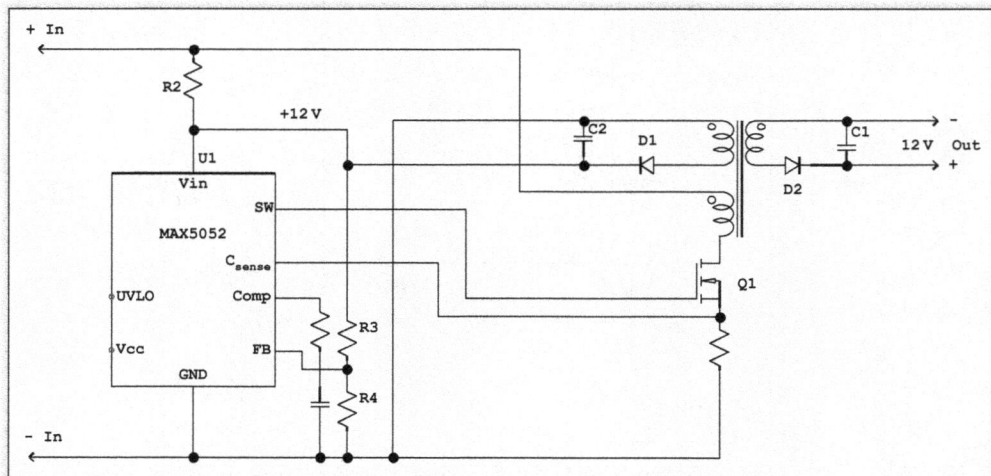

Figure 5-6: Feedback in a flyback converter using the auxiliary supply

auxiliary supply for the control IC. The feedback resistors (R3, R4) are chosen so that the control IC keeps the output voltage at 12.0 V. The filter capacitor (C2) on the IC supply adds a pole to the transfer function of the feedback loop, so the compensation becomes more complex. This control method is adequate for low power circuits where the regulation requirement is not too stringent. The voltage across D2 will vary with output current. As D2 drops more voltage, the output voltage will go down. The change in output voltage is not reflected in a change on the voltage of C2, so the regulation is no better than the output diode drop variation over the range of the output current.

A bootstrap circuit (R2, C2) is required to provide the initial voltage for the IC when using an auxiliary winding on the main transformer, as shown in Figure 5-6. All five transformer circuits can take advantage of a bootstrap circuit in conjunction with powering the IC from the main transformer. The bootstrap circuit will work with any control IC that has an under-voltage lockout circuit with hysteresis. The bootstrap resistor will charge up the IC supply capacitor slowly until it reaches the under-voltage enable voltage. The capacitor must store enough energy to drive the IC and the switch for a few cycles until the main power supply can supply all of the current required by the IC and switch drive. The bootstrap resistor supplies the charge current as long as AC power is supplied. This causes both heat and a reduction of efficiency. The advantage is that the resistor is an inexpensive part that is also very small compared to an iron core transformer as used in Figure 5-2. The bootstrap circuit is an excellent implementation for universal input supplies. ST produces a line of control ICs with the trade name VIPer that integrate the bootstrap circuit, as well as a high voltage MOSFET switch for very low part count operation in low power applications. National, Linear Technology, and other manufacturers also produce low power (under 20 W) fully integrated flyback circuits that require only a transformer and a few rectifiers and capacitors.

Flyback Circuits

A flyback converter works in a fashion similar to a boost converter, where energy is stored in the inductor while the switch is on and the energy is delivered to the load when the switch turns off.

Magnetic cores do not store magnetic energy very well. Efficient cores saturate at a low magnetizing force. A flyback circuit actually stores the energy of the inductor in an air gap. The core provides a low reluctance shielded path to couple the energy from the windings to the air gap. The energy storage is concentrated in the gap between the core faces.

Figure 5-7 shows a stylized ferrite core with the three windings for the circuit in Figure 5-6. The magnetic core concentrates almost all of the flux in the magnetic circuit inside the magnetic material. In a real core, there will be a very small amount of flux outside of the core in the vicinity of the windings, but all three windings will have essentially identical flux.

Recall the two equations for the voltage across an inductor:

$$V = L \; di/dt \text{ and } V = N \; d\Phi/dt$$

When the switch in Figure 5-6 closes, the current and flux will begin changing in proportion to the applied voltage on the primary inductor. The change in flux creates a voltage on each secondary winding in proportion to the turns of each winding. Since the voltage induced is negative (notice the dots on the windings), the diodes will not allow current to flow. When the switch opens, $d\Phi/dt$

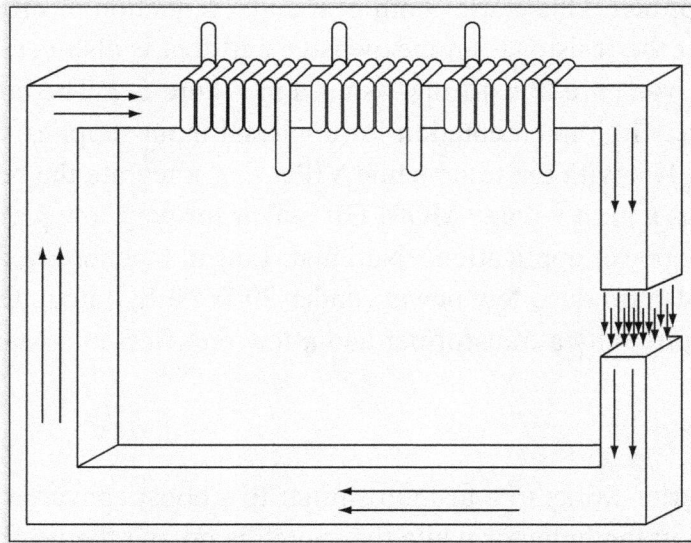

Figure 5-7: A stylized ferrite core with the three windings for the circuit in Figure 5-6

will change polarity instantaneously. As soon as N $d\Phi/dt$ is large enough to produce a voltage sufficient to forward bias one of the diodes, current will begin to flow in that secondary circuit. The consequence of this is that the secondary circuit with the lowest V/N ratio will hog all of the current from the collapsing magnetic field. Once the V/N ratio is equal for all secondary circuits, each will receive current from the collapsing field. This current hogging by the lowest V/N circuit is responsible for the close regulation of output voltage between all the secondary circuits. This is also why we can use the voltage on a secondary winding as a proxy for the voltage on the main output supply, as described above.

A flyback circuit can operate in either continuous or discontinuous mode. In continuous mode, current is always flowing in one of the windings of the inductor. In discontinuous mode, the current in all windings goes to zero for part of the cycle and the energy stored in the inductor goes to zero. Each mode has its advantages and disadvantages.

The primary advantage of continuous mode operation is that the relatively long current flow in the secondary requires a small filter capacitor (with larger allowable ESR). The primary inductance is relatively large with a small peak current requirement, so the inductance is relatively easy to implement. The peak current in continuous mode is roughly one-half that of discontinuous mode at the same power level. The principal disadvantage is that the control loop has a right-half plane zero that makes loop compensation difficult. However, the loop gain does not depend on the load current. It is only a factor of duty cycle and input voltage. Current mode controllers must also deal with slope compensation issues for continuous mode operation and duty cycle greater than 50%. Turn-on power dissipation in the switch is significant in continuous mode because the switch passes a large current as soon as the switch turns on with a large voltage applied. Another turn-on problem occurs because of the reverse recovery current in the output rectifiers. The reverse recovery causes an additional current spike during turn-on. Figure 5-8 shows representative waveforms for the circuit in Figure 5-6 when operated in continuous mode.

The discontinuous mode circuit trades many simplifications for larger peak currents. The turn-on dissipation in the switch is negligible because the current

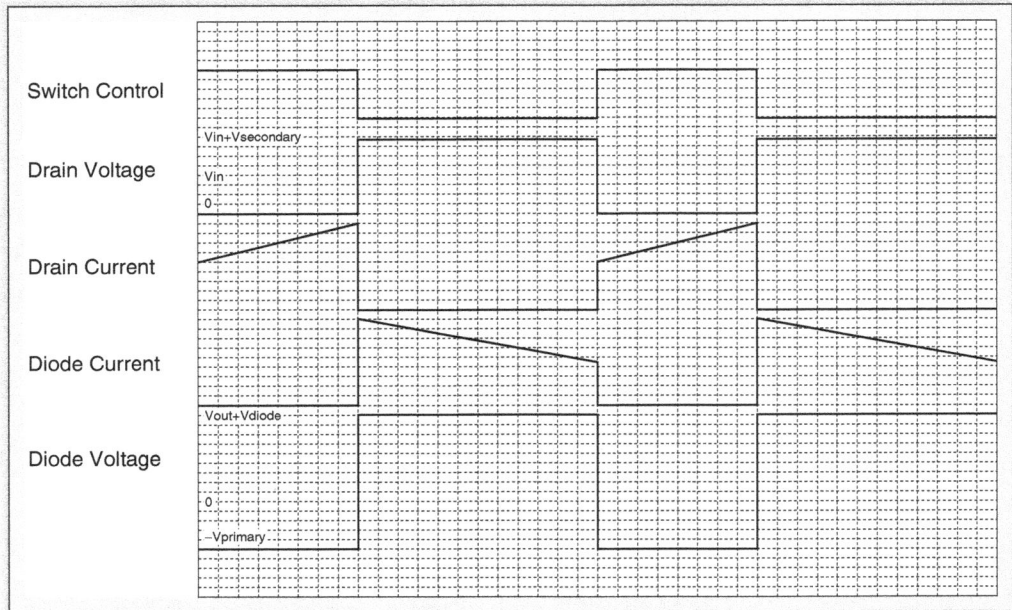

Figure 5-8: Representative waveforms for the circuit in figure 5-8 when operated in continuous mode

begins from zero and only the input voltage is applied to the switch. The output current goes to zero for part of the cycle, so there is no diode reverse recovery current to affect the switch during turn-on. The control loop is relatively straightforward in discontinuous mode. There is no right-half pole to deal with and slope compensation is never required. However, the load resistance is one of the factors in the loop equation. This makes open loop behavior less controlled than the continuous mode case. This is usually not a problem once proper compensation is accomplished and the loop is closed. The size of the gap in the inductor core becomes an issue for discontinuous mode because the higher peak current is likely to push the core closer to saturation. The AC flux in the core is quite large, so loss in the core is also an issue for discontinuous mode. The output ripple is typically larger in discontinuous mode because the AC current in the capacitor ESR is larger and the capacitor must supply the entire load current for a longer portion of the switching cycle. The simplicity in design, repeatability, and compensation makes discontinuous mode preferable, especially for low power circuits. Figure 5-9 shows representative waveforms for discontinuous operation. Discontinuous mode also has faster transient response and lack of load dump concerns as compared to continuous mode operation.

Switching circuits have parasitic inductances that are not associated with the energy storage inductor. These inductances are due to circuit traces and the leakage inductance of the main inductor. The parasitic inductances create a voltage that adds to the primary winding voltage, so the switch breakdown must be larger than the voltage implied by the reverse voltage plus input voltage. The turn-on time for the output diodes creates a short period of high secondary voltage, so there is a short time where *di/dt* becomes quite large. The extra *di/dt* from diode turn-on creates a spike on the primary.

Transformers and diodes have parasitic capacitances that can have undesirable consequences. The secondary capacitances, along with the secondary leakage inductance, can form a high frequency resonant circuit that is excited when the

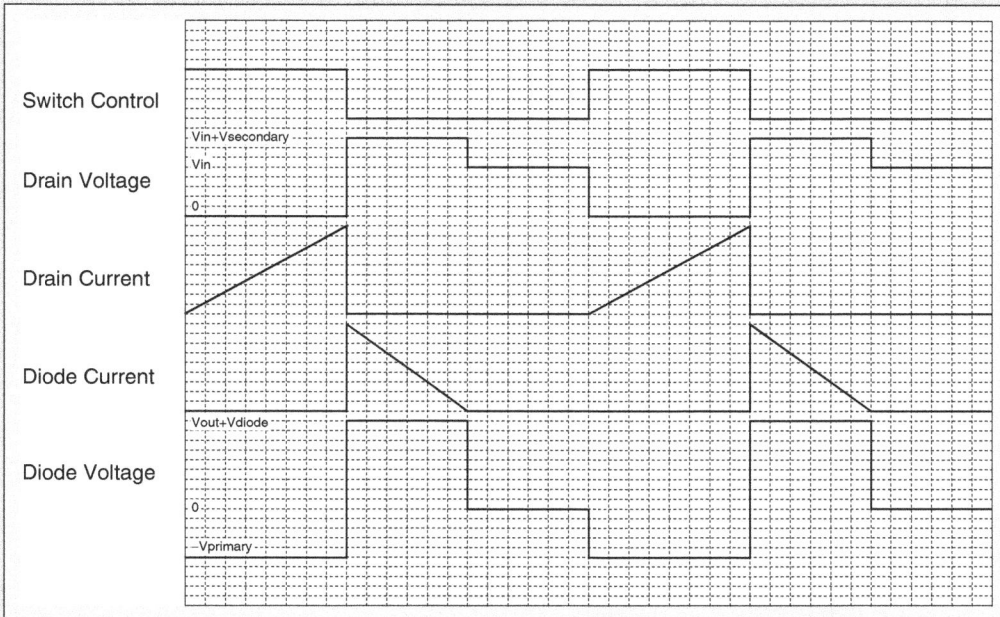

Figure 5-9: Representative waveforms for the circuit in Figure 5-6 when operated in discontinuous mode

diode turns off. This effect is more pronounced with hard recovery diodes. The resonant circuit will ring and transfer the AC waveform back to the primary.

Clamp circuits are used to reduce the stress on the switch from parasitic inductance elements. Figure 5-10 shows clamp circuits that will limit the voltage at the switch. Circuit A shows a clamp winding that returns energy from the magnetizing inductance of a transformer to the input supply. The clamp winding has the same number of turns as the primary. This sets the maximum voltage on the switch to twice the input voltage. Notice that D1 is connected to the input supply rather than between the clamp winding and ground. This topology

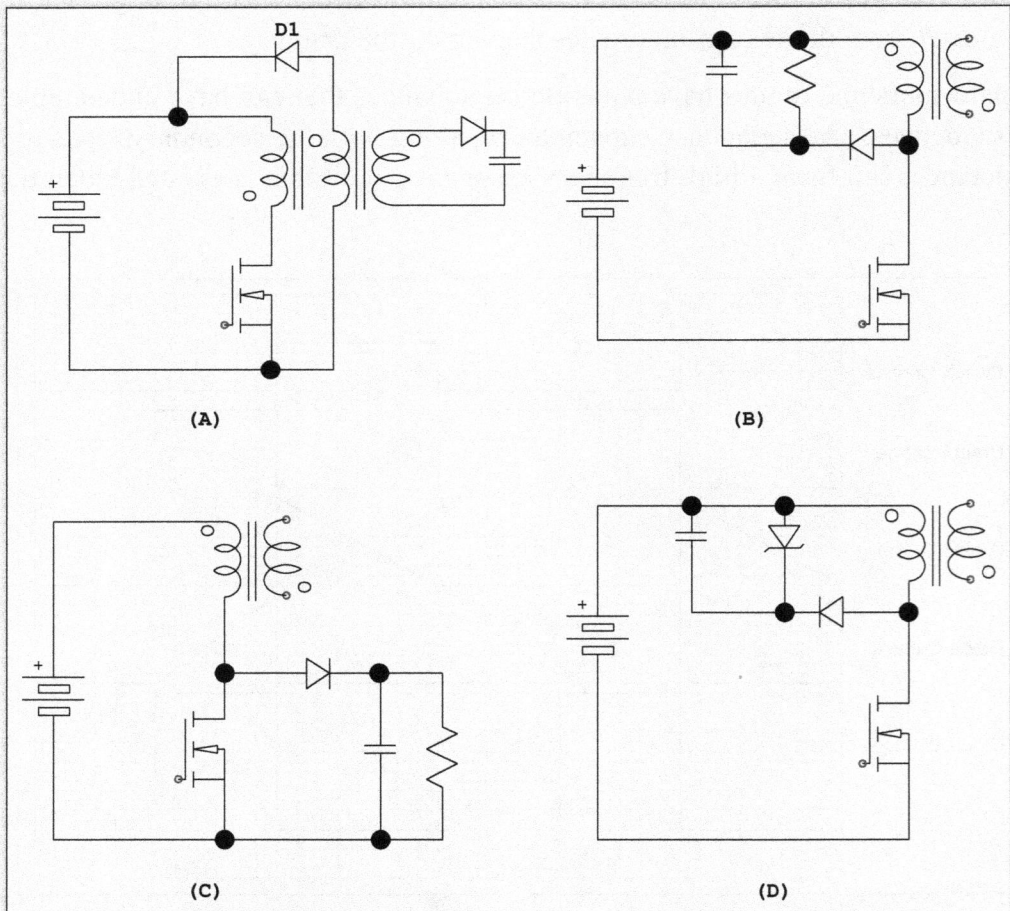

Figure 5-10: Clamp circuits that limit the voltage at the switch

is important because of the capacitances between windings. Placing the diode between the winding and ground will cause the capacitance to interfere with turn-on of the switch.

Circuit B uses the voltage of the capacitor to clamp the voltage at the switch. The time constant of the RC circuit is set to several switching cycles. The capacitor charges up to the reverse voltage created by the secondary windings plus any voltage from leakage inductances. This circuit is less efficient than the clamp winding because all of the energy stored in the leakage inductance and some of the energy of the primary inductance is dissipated in the resistor. Circuits C and D are variations of circuit B. The capacitor across the zener in circuit D may be necessary because zeners are not fast turn-on devices. The zener voltage must be set to a value larger than the normal voltage across the primary from secondary current flow.

Snubber circuits are similar to the clamp circuits. Figure 5-11 shows representative snubber circuits. The only interesting snubber circuits are those that dissipate energy in a resistor. Circuit A shows a simple RC snubber used on the output diode to dampen ringing when the diode turns off. The capacitor must be a small value so that the snubber provides a low impedance at the ringing frequency but a high impedance at the switching frequency. An RC snubber can also be used to slow both the rise and fall, as shown in Circuit B. This circuit will dissipate energy on both edges of the switching waveform. Circuit C shows a rate of rise snubber that limits the rate of voltage rise at the switch during turn-off. Circuit C is used to keep the voltage at the drain or collector of the switch low so that the power during the transition is kept low. The capacitor

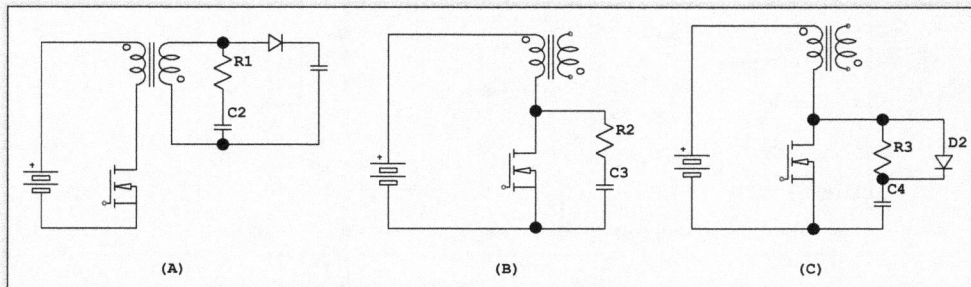

Figure 5-11: Representative snubber circuits

127

must be charged and discharged each cycle for proper operation. This requires a fairly low value for the resistor. A good rule of thumb is to set the RC time constant to 10% of the cycle time.

All of the protection circuits require fast turn-on diodes that have high peak current capability. The capacitor should have low ESR and low inductance to handle the high peak currents necessary. Ceramic and film capacitors are the preferred types. The resistor should have very low inductance. Wire-wound resistors should be avoided. The layout of the clamp circuit should avoid stray inductances so that the circuit does not create a new source of ringing and over-shoot. We will look at these protection circuits again in Chapter 7 when we look at the details of switch operation.

Figure 5-12 shows a two-switch circuit for a flyback circuit that allows use of lower voltage switches. The two diodes (D1, D2) clamp the primary winding to the input supply rails. This allows us to use switches with breakdown voltage just above the input voltage. The clamp action is efficient since the energy is returned to the input supply. T1 is the price we pay for using lower voltage switches. The transformer and resistors provide the necessary floating drive to Q1. The transformer drives both transistors to ensure that the switching times are as equal as possible.

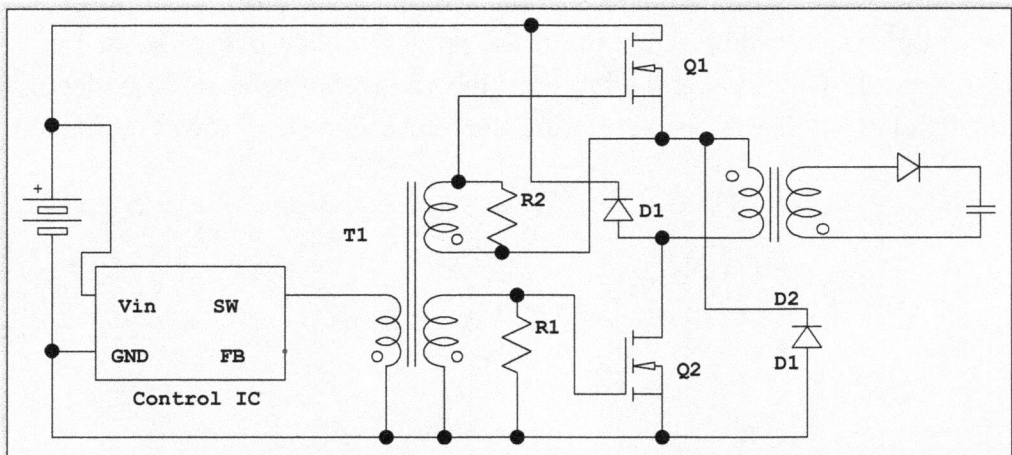

Figure 5-12: A two-switch circuit for a flyback circuit that allows use of lower voltage switches

Practical Flyback Circuit Design

Flyback design is usually iterative. We make educated guesses when picking component values and refine these in later iterations. Steps for designing a flyback circuit are listed below:

1. Choose a controller IC based on power level and bill of material constraints.

2. Choose the switching frequency.

3. Choose continuous mode or discontinuous mode.

4. Use the input voltage range to select the maximum duty cycle goal.

5. Determine the maximum power and pick a switch.

6. Design the primary inductance.

7. Design the transformer winding ratios.

8. Verify that the switch is adequate, based on the worst-case voltage.

9. Choose the bootstrap capacitor based on the gate charge required if a bootstrap supply is used.

10. Choose the output capacitor, based on ripple requirements.

11. Design the ancillary IC components.

Off-Line Flyback Example

Our first example is a universal input flyback design that has a 12.0 V/1 A output. The output must have regulation of ± 200 mV with 100 mV or less ripple. This is a circuit similar to the power supply for a number of consumer devices with an input specification of 100–240 VAC and output of 12 V/400 mA. Instead of a "wall wart" with an iron core transformer, these products integrate the entire switching power supply and power plug into a plastic housing about four times larger than a standard U.S. two-prong power plug. Figure 5-13 shows the circuit we are designing.

My first step for the example was to check the websites of Maxim, TI, and Linear Technology to see which controller ICs came up on a search for the

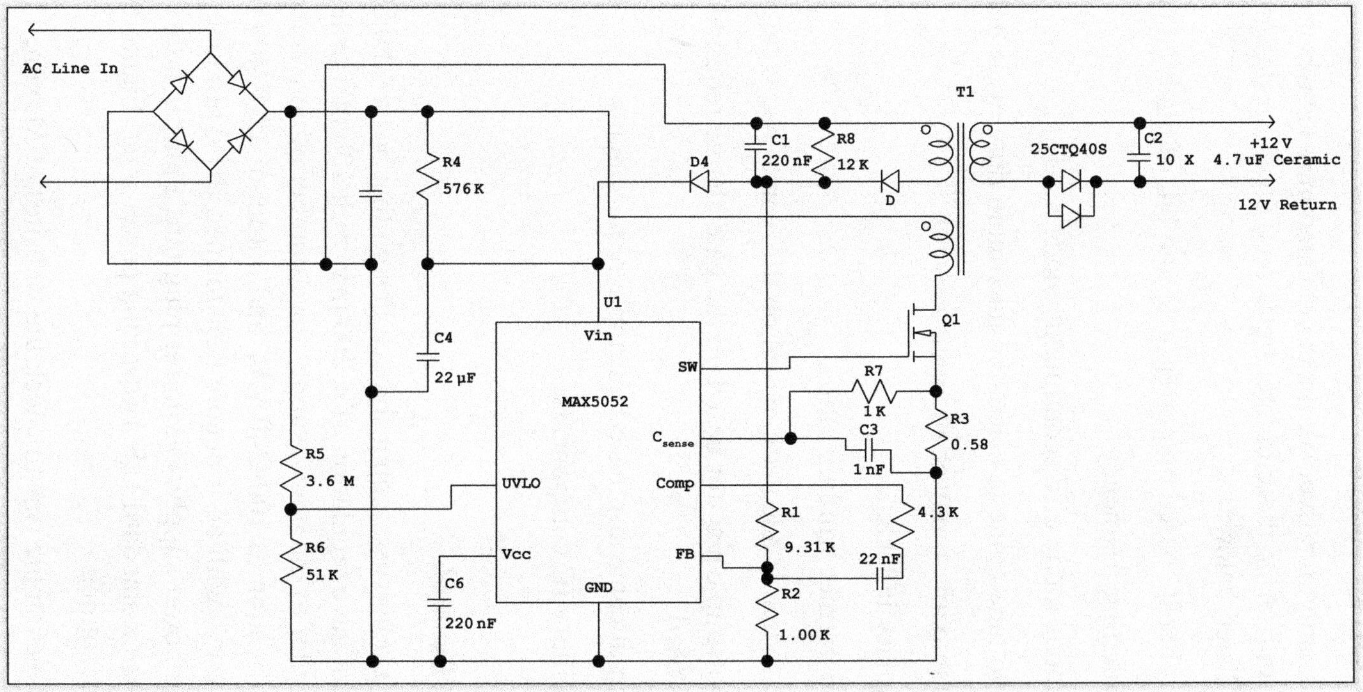

Figure 5-13: 12 V isolated flyback power supply using the MAX5052

"flyback" keyword. The Maxim_NPP_PWM_Products.pdf Application Note came up as a match. I found the MAX5052 that is designed for precisely the type of circuit we will design. This IC is designed for low to moderate power operation in a universal input application. Its major advantage is a very large hysteresis in the under-voltage lockout circuit. The worst-case hysteresis is 9.25 V and typical is 11.86 V. The large difference between the wakeup level and the shutdown level means that we can use a smaller reservoir capacitor and lower wattage resistor for the IC bootstrap power supply. This IC has a fixed switching frequency of 262 kHz, which should be appropriate for our design. The cycle time is 3.82 µs.

Discontinuous operation seems to be a reasonable choice, since we are aiming for a simple design. We have two possible choices for maximum duty cycle. The MAX5052A has a maximum duty cycle of 50% and the MAX5052B has a maximum duty cycle of 75%. Our initial choice is to limit the duty cycle to 50%.

The voltage applied to the main output rectifier and bias supply rectifier is likely to be close to double the output voltage plus the forward voltage of the diode. The main output voltage is only 12.0 V, so the PRV rating of the main rectifier can be on the order of 40 V. This allows us to use a Schottky diode for minimum power dissipation. The peak-to-average ratio of current for discontinuous mode can be large, so it is likely that peak rectifier current may be on the order of 10 A. An IRF 30BQ040 Schottky diode has a 40 V PRV rating and 3.0 A average forward current. Looking at the forward voltage versus instantaneous forward current shows that the voltage drop changes from 0.8 V at 10 A to only 0.25 V at 100 mA, so we will have a problem meeting the regulation specification.

We have a couple of options that we can consider at this point to improve the voltage regulation over input voltage and output current. The first option is to look for a diode with better voltage drop at high current; the second option is to change course and go to a continuous mode converter to reduce the peak output current. In searching the IRF website we find that a 6CWQ03FN dual diode has less change in forward voltage versus current. Additionally, each diode will only carry one-half the total current, so we stay in the more vertical part of the curve. This diode seems to have more than adequate ratings with

3.5 A per diode and 30 V PRV. It is also a relatively small surface mount component. Worst-case forward voltage is 0.5 V at 5 A and 25°C. This should barely meet the regulation requirement.

It is reasonable to also use 12.0 V for the auxiliary supply. This allows the secondary windings to be identical. The IC draws 2.5 mA maximum, so the IC will consume 12.5 V * 2.5 mA = 30 mW. A guess for the power to drive the MOSFET switch is 70 mW (an educated guess of twice IC power plus a little more). The power for the main output will be approximately 12.5 V * 1.0 A = 12.5 W. This means the total power will be 12.6 W.

The highest switch current will occur at the lowest input voltage (85 VAC or 115 VDC with 10 V of ripple). The average current is 12.6 W/115 V = 110 mA. If we make a guess of a 10:1 peak-to-average ratio, the peak input current will be 1.1 A. A search of the IRF website indicates that the IRBF20S MOSFET has 900 V breakdown and 1.7 A average current. This should be a good choice for the switch. Now we can calculate the drive power for the switch. The switch is driven by a regulated 10.5 V supply from the control IC. The switch drive current is the total gate charge times frequency or 38 nC * 262 kHz = 10 mA. The drive power is 10 mA * 10.5 V = 105 mW. Our guess was close enough to approximate the actual power.

There is a trade-off between duty cycle, turns ratio, primary inductance, and switch voltage. Longer duty cycle will require a larger primary inductance, but it will allow a smaller turns ratio and lower switch voltage. We can choose the primary inductance so that the circuit is at the crossover between continuous and discontinuous operation at 50% duty cycle and the lowest input voltage. We know that we need 12.6 W of power delivered to the load during one-half of a cycle. The average current during the time the switch is on is one-half the peak current because of the triangular shape of the current waveform (refer to the waveforms for discontinuous operation in Figure 5-9).

The average current for the whole cycle is I_{Peak} * 0.5 * Duty Cycle. From this, we calculate that I_{Peak} is 110/(0.5 * 0.5) = 440 mA.

We can use the inductor equation to calculate the primary inductance:

$$L = \frac{V * \Delta t}{\Delta I} = \frac{110 * (3.82\ \mu s * 0.5)}{440\ mA} = 478\ \mu H$$

The energy stored in the core is transferred to the output when the switch opens. Flyback designs have an additional element of freedom compared to boost designs, since the inductance of the secondary and output voltage will set both peak current and *di/dt*. In discontinuous operation, we know that all of the energy will be transferred to the output circuit before the switch closes again. This puts an upper limit on *dt*. The largest secondary inductance possible will be a value that puts *dt* equal to (1 − Duty Cycle). We can make the secondary inductance smaller, if desired. The ratio of primary to secondary inductance sets the turns ratio of the inductor. Smaller secondary inductance gives a larger turns ratio that will also create a larger voltage requirement for the switch.

We will choose Δt equal to one-half of the cycle time. Now we can choose the secondary inductance. We use the highest current output to select the secondary inductance. We know that the main output will need 12.5 V at 1.0 A. The output waveform is a triangle as shown in Figure 5-9, so the average output current is $I_{Peak} * 0.5 * (1 - \text{Duty Cycle})$. From this, we can calculate that I_{Peak} is $1.0/(0.5 * 0.5) = 4.0$ A. Once again, we use the inductor equation to calculate the secondary inductance:

$$L = \frac{V * \Delta t}{\Delta I} = \frac{12.5 * (3.82\ \mu s * 0.5)}{4.0A} = 5.96\ \mu H$$

We can calculate the turns ratio from the two inductances. The equation for inductance is

$$L = N^2 * A_L,$$

so we can use this equation to develop the turns ratio in terms of inductance ratios:

$$\frac{L_P}{L_S} = \frac{N_P^2}{N_S^2}, \text{and rearranging gives } \frac{N_P}{N_S} = \sqrt{\frac{L_P}{L_S}} = \sqrt{\frac{456}{5.96}} = 8.75{:}1.$$

Going back to one of our inductor equations, $V = N\, d\Phi/dt$, and recognizing that $d\Phi/dt$ is identical for all windings, we get the relationship of the voltages during switch on-time and switch off-time.

$$\frac{V}{N} = \frac{d\Phi}{dt}, \text{so } \frac{V_P}{N_P} = \frac{V_S}{N_S} \text{ or } \frac{V_P}{V_S} = \frac{N_P}{N_S}.$$

This appears to be the same as the transformer equation. It is similar, since the windings are coupled, but it is important to remember that it is the inductances of the windings that set the voltages. The transformer equations only apply when current is flowing in the primary and secondary at the same time.

During switch on-time the reverse secondary voltage is controlled by the input voltage, so the worst-case PRV on the secondary diode will be:

$$V_S = \frac{N_S * V_P}{N_P} = \frac{1 * 390}{8.75} = 45 \text{ V}.$$

The worst-case switch voltage while the switch is off will equal the highest input voltage plus the reverse voltage on the primary:

$$390 + (12.5 * 8.75) = 390 + 110 = 500 \text{ V}.$$

We see that our choice for the secondary will over stress the output diodes and place minimal stress on the switch. We can shorten the output current time by reducing the secondary inductance and increase the turns ratio. We can increase the turns ratio by 33% and see if the diode and switch characteristics are more reasonable. We set the turns ratio to 12.0. This yields a secondary inductance of 3.17 μH. The PRV for the diode becomes $390/12 = 32.5$ V. The peak output current will be 7.5 A. The worst-case switch voltage will be 540 V. We need another iteration of choosing the output diode. A search of the IRF website yields the 25CTQ40S, which is in the same package as the 6CWQ03FN. The 25CTQ40S dual diode has even better forward voltage characteristics and has enough margin with 40 V PRV.

The typical values for wakeup voltage (21.6) and shutdown voltage (9.74) give a change of 11.86 V. However, the worst-case operation of the IC bootstrap occurs when the IC wakes up at the lowest voltage and shuts down at the highest voltage. The lowest wakeup is 19.68 V; the highest shutdown voltage is 10.43 V. The current draw is relatively constant. The IC draws 2.5 mA and the gate charge draws an additional 10 mA. We allow 10 ms for the supply to charge the auxiliary bias supply above 10.43 V. 12.5 mA for 10 ms means we will use 125 μC of charge. We can use the capacitance equation $Q = C * V$ and the change in charge to obtain an equation for capacitance:

$$Q_2 - Q_1 = 125 \ \mu C - C*V_1$$

$$C4 = \frac{Q_2 - Q_1}{V_2 - V_1} = \frac{125 \ \mu C}{19.68 - 10.43} = 13.5 \ \mu F.$$

Rounding to the nearest value $= 22 \ \mu F$.

This capacitance will be necessary because the bias supply will get no current from the switch until the main output voltage equals the bootstrap voltage (when both windings have equal V/N). The bootstrap resistance value (R4) is a compromise between fast startup and power dissipation. We can limit the power dissipation to 0.25 W to keep heat low and maintain high efficiency. The worst-case voltage is 390 V − 12.0 V = 378 V. The resistance required is $378^2/0.25$ W = 571 k. The bootstrap charging current is 378 V/571 k = 660 μA. The nominal charge to reach the wakeup point is 22 μF * 20 V = 440 μC, so it will take 0.67 seconds to charge the bootstrap capacitor at high input (240 VAC) and 2.6 seconds at low input (100 VAC).

The next step is to choose the output capacitor. We are likely to encounter the same problem we saw with non-isolated circuits in Chapter 4, where the capacitance value is secondary to ESR in setting output ripple. Our goal is to assign 67% of ripple voltage to ESR and 33% to AC impedance, so we assign 67 mV of ripple to ESR.

$$ESR = \frac{67 \, mV}{7.5 \, A} = 8.9 \ m\Omega$$

The target capacitance is :

$$X_C = \frac{33 \, V}{7.5 \, A} = 4.4 \ m\Omega$$

$$C = \frac{1}{2 * \pi * 262 \, kHz * 4.4 \ m\Omega} = 140 \ \mu F$$

A quick look in the Digi-Key catalog shows the Panasonic CD series polymer electrolytic would require seven 8.2 μF/16 WV capacitors to have low enough ESR and enough ripple capability. A search of the Panasonic website shows a 4.7 μF/16 WV MLC ceramic can handle 4 A of current and each capacitor has

9 mΩ for ESR. In the case of the ceramic capacitors, we will need multiple capacitors to have enough capacitance, and the ESR becomes quite small. Ten of these capacitors will probably make a better selection than the electrolytic capacitor. This would give only 0.9 mΩ ESR. This reduces the required capacitance to 45 µF. An aluminum electrolytic capacitor will be sufficient for the bias supply, since the total current is only 13 mA.

Notice that the bias supply has two stages of filtering isolated by diode D4. This allows the voltage at the feedback pin to follow the output voltage during startup so that the internal soft start circuit is not affected by the voltage from the bootstrap circuit. The time constant of the resistor in parallel with the feedback capacitor in the feedback portion is rather short (on the order of three cycle times). This allows the feedback to more closely follow a drop in the main output voltage.

The feedback voltage divider is calculated from the equation given in the data sheet:

$$V_{OUT} = \left(1 + \frac{R_1}{R_2}\right) \times 1.23 \text{ V}$$

The current sense resistor is calculated based on the worst-case peak current required. We calculated that the peak current in normal operation at 85 VAC input is 440 mA. We can set the current limit to a value slightly above this value to allow for additional current during startup. We choose 500 mA, so

$$R_{CS} = \frac{0.29 \text{ V}}{0.5 \text{ A}} = 0.58 \text{ }\Omega.$$

We add a small amount of RC filtering (R7, C3) between the current sense resistor and current sense pin to allow for some transients when the switch turns on. This reduces false current limiting due to transients. The value of the capacitance can be adjusted in the lab. It is possible that this capacitor may not be needed.

The IC driver can sink and source more than 650 mA, so there is no need for a current limit between the gate of the switch and the IC.

The compensation components are taken from the data sheet. They serve only as a starting point. Actual compensation will need to be adjusted in the laboratory to ensure a stable loop.

Resistors R5 and R6 set the under-voltage lockout value. The voltage at this pin must be 1.28 V before the IC will operate. A reasonable input voltage is 95 V for this pin to be active. The value of R5 is very large, so the bias current of the UVLO pin will affect the value needed for R6. We can consider V_{IN} and R5 to be a constant current source, so we need to subtract the bias current from the current supplied by R5 when calculating R6. The data sheet also gives equations for calculating these resistors.

Non-Isolated Flyback Example

Our next example shows the advantage of a non-isolated flyback design for automotive use. An automotive system can range from 11.5 V at low battery with the key off to 15.0 V when charging a drained battery. Some systems are designed to work at a nominal 13.6 V ± 0.5 V. This represents full voltage for a charged battery. Our example implements a system that produces 13.6 V at 10 A. The output ripple target is 300 mV. The regulation target is 400 mV. Figure 5-14 shows our circuit.

A reasonable choice for the control IC is the LT1680. This IC is designed for high power step-up DC–DC converters using an external MOSFET switch. It provides all of the necessary current mode PWM functions and will operate directly from the input supply.

A reasonable switching frequency is 167 kHz. The maximum frequency of the IC is 200 kHZ, but we want to stay away from effects where we have no control. The cycle time is 6.0 μs. This frequency is low enough that parasitic effects at the high power level will be manageable. This frequency is also within the power range of reasonably priced inductor cores.

Continuous mode operation is a reasonable selection for this design. The output current will approximate the input current, since the input voltage range is +10/−20% of the output voltage range. Choosing continuous mode will allow the peak current to be only slightly larger than twice the output current. If we set the 50% duty cycle voltage to 10.5 V input, we will have enough margin when the voltage drops to 11.0 V to maintain control and avoid the need for slope compensation. This sets the target duty cycle for the lowest input voltage around 40%, as a first guess. We use the graph in the data sheet to choose the

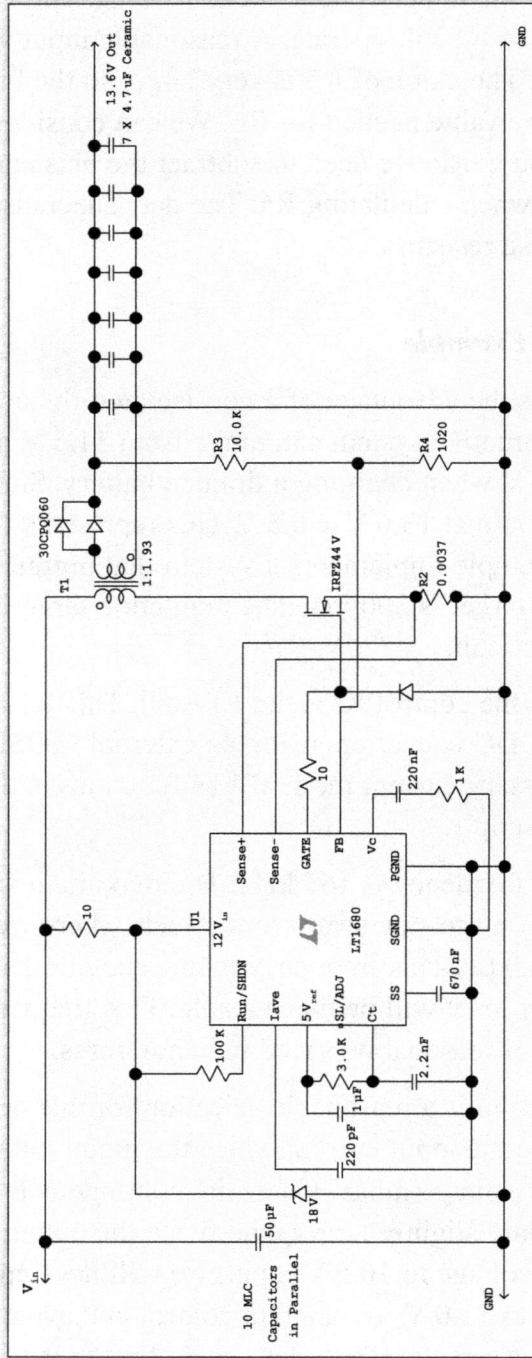

Figure 5-14: 13.6 V non-isolated flyback supply for automotive systems using the LT1680

3 K timing resistor based on our maximum duty cycle. Another graph in the data sheet gives us 2.2 nF based on the 167 kHz frequency and 3 K timing resistor.

A 60 V Schottky diode is a reasonable first try at the output rectifier. The turns ratio of the inductor is likely to be very close to 1:1. It is a reasonable guess that the turns ratio will be no larger than 1:2. The IRF 30CPQ060 has 60 V PRV and 30 A average current and is a dual diode package. The peak forward current is likely to be approximately 20 A, so this diode should fit our requirements. Each diode will pass one-half of the total current, so the forward voltage drop will be 0.55 V. The maximum output power will be 13.6 V * 10.0 A + 0.55 V * 10.0 A = 141.5 W.

The worst-case switch current will occur at 11.0 V input and the worst-case switch voltage will occur at 15.0 V input. A good rule of thumb for the switch voltage is to assume it will be twice the highest input voltage. Another rule of thumb is to pick the switch current equal to twice the average current plus the ripple factor. We will pick the ripple current equal to 30% of average current to allow a reasonable amount of dynamic response. This low ripple factor also allows a larger amount of ESR in the output capacitor. The average input current will be:

$$\frac{1}{DC} \times \frac{\text{Load Power}}{\text{Input Voltage}} = (1/0.4)*(141.5 \text{ W}/11.0 \text{ V}) = 32.2 \text{ A}$$

The peak primary current will be 32.2 * 1.15 = 37.0 A (input current times ripple factor). The ripple current will be 32.2 * 0.3 = 9.66 A. The IRFZ44V that we used in Chapter 4 is a good choice for this application, too. It has 60 V V_{DSS} and 55 A I_D.

Now we can start designing the primary inductance. We have constrained the primary inductance by the expected ripple current, the duty cycle, and the input voltage. We use the rearranged inductor equation again:

$$L = V \frac{dt}{dI} = 11.0 * = \frac{0.4 \times 6 \text{ } \mu s}{9.66 \text{ A}} = 2.7 \text{ } \mu H$$

From Chapter 1, we recall the formula for flyback operation in continuous mode:

$$V_{OUT} = V_{IN} * N * \frac{DC}{1 - DC}$$

We can determine the turns ratio (secondary turns/primary turns) from our starting assumptions:

$$N = \frac{V_{OUT} \times (1 - DC)}{V_{IN} \times DC} = \frac{(13.6 + 0.55) \times (1 - 0.4)}{11.0 \times 0.4} = 1.93:1$$

The worst-case switch voltage is high input plus the reflected secondary voltage: $15.0 + (14.15 * (1/1.93)) = 22.3$ V. The switch has more than enough head room so that a clamp circuit is probably not necessary to protect the switch. The worst-case power dissipation for the switch is peak current squared times on-resistance times duty cycle: $(37 \text{ A} * 37 \text{ A}) * 0.016 \, \Omega * 0.4 = 8.8$ W. Actual power dissipation will be slightly higher once we take switching losses in to account. The worst case for the rectifier is high input voltage times turns ratio: $15.0 \text{ V} * 1.93 = 29.0$ V. The average diode current when the switch is off is output current divided by $(1 - DC)$: $10A * 0.6 = 16.7$ A. The peak output current is peak input current times the turns ratio: $37.0 \text{ A} * (1/1.93) = 19.2$ A. These calculations show our choice of semiconductors is appropriate.

We again assign 67% of ripple voltage to the ESR of the output capacitor, so:

$$ESR = \frac{200 \text{ mV}}{19.2 \text{ A}} = 10.4 \text{ m}\Omega$$

$$X_C = \frac{100 \text{ mV}}{19.2 \text{ A}} = 5.2 \text{ m}\Omega$$

$$C = \frac{1}{2 * \pi * 167 \text{ kHz} * 5.2 \text{ m}\Omega} = 180 \text{ }\mu F$$

This value is similar to our previous example and will require multiple ceramic or aluminum capacitors to satisfy both the ESR requirement and the ripple current requirement. The capacitors for the previous example will be adequate for this design when enough are used in parallel. The most significant requirement for the capacitors is ripple current capability. Seven of the 4.7 µF/16 WV MLC capacitors will have only 1.3 Ω ESR, so the 33 µF combined capacitance will be more than enough to meet our ripple voltage requirement. Using so many capacitors in parallel will invite problems with EMI and secondary problems that will increase ripple unless we pay strict attention to proper layout. The connections to the capacitors should be made with very wide but closely spaced conductors. This will reduce the inductance of the traces and minimize the loop area of the traces.

The current sense resistor is set by average current rather than peak current for this control IC. The equation is found in the data sheet:

$$R_{CS} = 120\,mV/I_{AVG} = 0.12\,V/32.2\,A = 3.7\,m\Omega$$

The average current limit is set by the combination of the current sense resistor and the current limit integration capacitor. The data sheet recommends setting this capacitor to 220 pF.

The output voltage is set by the equation:

$$V_{OUT} = \left(1 + \frac{R_1}{R_2}\right) \times 1.25\,V$$

Running the calculations requires a resistor ratio of 9.88:1.

We can set the soft start time to 100 ms, using the equation from the data sheet:

$$C_{SS} = 0.1\,s/150,000 = 670\,nF$$

Again, we start with the compensation values from the data sheet and will change them based on results in the lab. No slope compensation is necessary, since duty cycle is limited to 50%.

The large current pulses on the input will require very low ESR to maintain the voltage at the control IC. Selecting input capacitors equal to the output capacitors will provide the necessary low ripple. The very large input current pulses may make a forward converter a better choice for this application.

Forward Converter Circuits

A forward converter is a single switch converter that uses a transformer to transfer energy from the primary circuit to the secondary circuits. Energy flows from the primary to the secondary while the switch is conducting current. Figure 5-15 shows a representative circuit for a forward converter. A voltage clamp is necessary for a forward converter because all transformer current stops when the switch turns off. The clamp provides a path for the current in the magnetizing inductance of the transformer and the leakage inductance. In the flyback circuit, the current flow in the secondary provides a path for the flux of the core when the switch opens; the clamp is only necessary to reduce stress on the switch from leakage inductances.

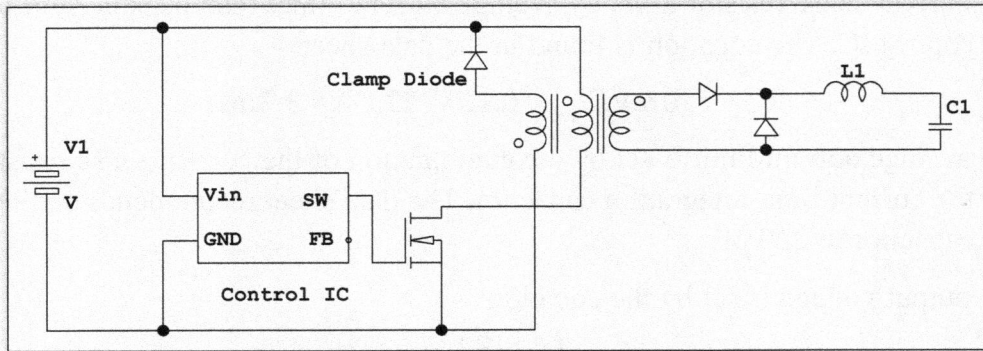

Figure 5-15: Representative circuit for a forward converter

Any of the clamp circuits in Figure 5-10 can be applied to the forward converter. The clamp circuits will have a voltage that is controlled by the secondary voltage when used in a flyback circuit because of the requirement that V/N is equal for all windings. This is not true for the forward converter. The clamp winding in Figure 5-10(a) guarantees that the switch voltage is twice the input voltage while the magnetizing current ramps down. Circuits B and C will have varying voltages depending on the amount of energy that is dissipated in the resistor. You must exercise care in designing the maximum duty cycle, transformer magnetizing inductance, and RC time constant when using Circuits B and C to ensure you do not exceed the voltage rating of the switch. Notice that circuit C is identical to a boost regulator. International Rectifier Application Note AN-939A gives a very good description of using dissipative clamp circuits in forward converters.

The clamp circuit design affects the maximum switch voltage required for a forward converter. The energy stored in the magnetizing inductance is proportional to the volt-seconds while the switch is on. The same number of volt-seconds is necessary to dissipate the energy stored in the magnetizing inductance of the transformer during the switch off-time. The voltage stress on the switch can be reduced by limiting the duty cycle. However, reducing the duty cycle will increase the primary peak current and the output peak current and voltage. The clamp winding generally has the same number of turns as the primary which sets the switch voltage to twice the input. However, the maximum duty cycle and clamp winding turns can be adjusted to set the switch voltage to any desired value. Our second example will show how to use a large switch voltage to reset the flux in the core when the duty cycle is greater than 50%. The clamp circuit only dissipates the energy in inductances inside the loop created by the clamp circuit. Any parasitic inductances outside the clamp circuit, such as switch lead inductances, will create voltages when the switch turns off and will add to the voltage stress on the switch.

The same two switch circuit we saw in Figure 5-12 can be used for a forward converter circuit by substituting a transformer for the flyback inductor. The maximum voltage on each switch will be slightly higher than the input voltage. The diodes again clamp the reverse voltage of the transformer inductance to the input voltage. Since the clamp voltage can be no larger than the input voltage, the duty cycle must be restricted to a value less than 50% to ensure that flux does not build up in the core and result in saturation.

Practical Forward Converter Design

Typical steps for designing a forward converter are listed below:

1. Choose a controller IC based on power level and bill of material constraints.

2. Choose the switching frequency.

3. Use the input voltage range and output ripple current goal to select the maximum duty cycle goal.

4. Pick the output diodes.

5. Design the transformer winding ratios.

6. Determine the maximum power and pick a switch.

7. Choose the bootstrap capacitor based on the gate charge required if a bootstrap supply is used.

8. Calculate the output inductor value.

9. Choose the output capacitor based on ripple requirements.

10. Design the auxiliary supply, if needed.

11. Design the ancillary IC components including the feedback circuit.

Off-Line Forward Converter Example

Our first example is a universal input off-line supply to provide 5.0 V at 20 A. (See Figure 5-16.) The ripple voltage is required to be below 100 mV and the regulation is required to be 200 mV. Even though the feature list of the MAX5052 says it is good for 50 W of output power, there is no reason it cannot be used above that power level as long as it is able to drive the switch. We will choose the MAX5052A for 50% maximum duty cycle. A 45% duty cycle is reasonable for the very lowest input voltage. This allows enough margin that the supply will start with the lowest input voltage for a 100 VAC power system. We will want to keep the output ripple current to a minimum to keep the ripple voltage low. We can choose an output ripple target of 10% or 2 A. The easiest clamp design is to use a winding on the power transformer and a diode (D3). D3 must be a fast-turn-on diode. The current through the diode will go to zero, so we are not concerned with turn-off characteristics.

There will be a constant diode drop in the output circuit because inductor current will flow for the whole switch cycle. Schottky diodes are the preferred components in low voltage supplies with modest power output. We can choose a dual diode that can handle the peak current. Our peak current is 20 A + 1 A ripple. The IRF 30CPQ060 is a dual diode in a TO-247AC package with a 30 A average current rating and 60 V PRV rating. This diode has a forward voltage of 0.7 V at 20 A forward current.

We use a rearranged version of the buck converter equation from Chapter 1 to determine the required input voltage.

$$V_{IN} = (V_{OUT} + V_{Diode})/DC = 5.7 \text{ V}/0.45 = 12.7 \text{ V}$$

This voltage must be present at the secondary winding at the lowest input voltage. This gives the turns ratio of the transformer:

$$N = 100 \text{ V}/12.7 \text{ V} = 7.9$$

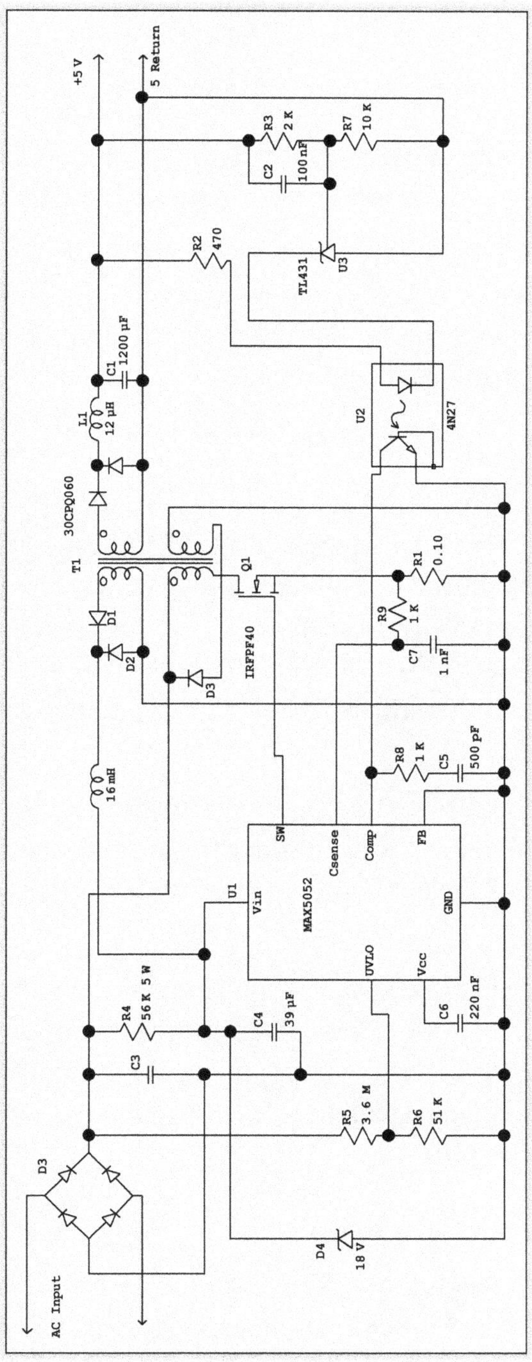

Figure 5-16: Representative universal input off-line forward converter supply

We can verify the required duty cycle at high input voltage. The input voltage will be:

$$390 \text{ V}/7.9 = 49.5 \text{ V}$$

This means the duty cycle at high input will be $5.7/49.5 = 11.5\%$. The high input voltage confirms that the diode we chose is adequate.

The power delivered must be $5.0 \text{ V} * 20 \text{ A} + 0.7 \text{ V} * 20 \text{ A} = 114 \text{ W}$. The maximum current in the switch will occur at low voltage. The average switch current is calculated from average power and peak current is calculated from average current, duty cycle, and ripple factor:

$$I_D = 114 \text{ W}/100 \text{ V} = 1.14 \text{ A}$$

$$I_{D\text{-Peak}} = 1.14 \text{ A}/0.45 * 1.05 = 2.7 \text{ A}$$

The switch will need more current capacity than this, once all the power consumption sources are factored (auxiliary supply, switch losses, transformer losses, inductor losses, capacitor losses, etc.). We will need a switch with a 900 V/5 A rating. The IRF IRFPF40 MOSFET has a 900 V V_{DS} and 4.7A I_D rating with 2.5 Ω R_{DSON}. The total gate charge is 120 nC, so the gate drive current is 120 nC * 262 kHz = 32 mA.

We are less worried about bootstrap power in this design, so we can allow more dissipation in the bootstrap resistor in order to keep the startup time short. A good rule of thumb is to have the system start within 500 ms at the lowest input voltage.

The current draw is relatively constant. The IC draws 2.5 mA and the gate charge draws an additional 32 mA. We allow 10 ms for the circuit to charge the bias supply above 10.43 V. We will use 345 µC to supply 34.5 mA for 10 ms charge. We again use the capacitance equation to calculate the required capacitance:

$$C_4 = \frac{Q_2 - Q_1}{V_2 - V_1} = \frac{345 \text{ µC}}{19.68 - 10.43} = 37 \text{ µF.}$$

Rounding to the nearest value gives 39 µF.

The nominal charge to reach the wakeup point is 39 µF * 20 V = 780 µC. This means we will need 1.6 mA to charge the capacitor in 500 ms. Subtracting the capacitor voltage from the input voltage and dividing by the required current gives 90 V/1.6 mA = 56 K. The maximum power will occur at high voltage, so $(390 - 20)^2/56$ K = 2.5 W. This resistor will need to be a 5 W resistor.

The inductor value is determined by the ripple current, applied voltage, and duty cycle. The applied voltage is the transformer voltage less the diode drop minus the output voltage. We apply the inductor equation:

$$L = V \frac{dt}{dI} = (12.0 - 5.0) * = \frac{0.45 \times 3.82 \, \mu s}{1.0 \, A} = 12.0 \, \mu H$$

The output capacitor value is determined by the ripple voltage requirement. We have 100 mV of ripple and 1.0 A. We can choose ESR and the capacitor value using our one-third and two-thirds rule:

$$ESR = \frac{67 \, mV}{1.0 \, A} = 67 \, m\Omega$$

The target capacitance is

$$X_C = \frac{33 \, mV}{1.0 \, A} = 33 \, m\Omega$$

$$C = \frac{1}{2 * \pi * 262 \, kHz * 33 \, m\Omega} = 18 \, \mu F$$

A good choice for the output capacitor is a Panasonic series FM Type A. There are no capacitors close in value to 18 μF in the 6.3 WV range. The closest value that has a low enough ESR and enough ripple capacity is the EEUFM0J122L 1200 μF capacitor that can handle 1.56 A of ripple and has 30 mΩ ESR.

The auxiliary supply needs to provide approximately 12 V for normal operation but must not go above 30 V. The diodes D1 and D2 can be small Schottky diodes with 60 PRV. The auxiliary supply is not regulated and there is no coupling between the main output and the auxiliary supply. It is very likely that the auxiliary supply will rise to a large voltage during startup and during large transients on the main output. The zener diode shunt (D4) is provided to ensure that extra current will keep the supply within the limits of the control IC. The zener voltage is set high enough that it normally will not draw current. We can choose a very low value for the inductor ripple because the current is essentially constant. There is no need for rapid transient response, and the low ripple will reduce fluctuations in output voltage during main output transients. We choose 5% ripple current for this supply, or 34.5 mA * 0.05 = 1.7 mA.

We calculate the inductor at the lowest input voltage. We also use lowest input voltage to calculate the turns ratio for this supply.

$$V_{IN} = (V_{OUT} + V_{Diode})/DC = 12.7\,V/0.45 = 28.2\,V$$

$$N = 100\,V/28.2\,V = 3.6$$

$$L = V\frac{dt}{dI} = (27.5 - 12.0) * \frac{0.45 \times 3.82\,\mu s}{1.7\,mA} = 16\,mH$$

The last step is to design the feedback circuit. We will use a standard 4N27 optoisolator and a TL431 shunt regulator to provide feedback to the control IC. We choose a small amount of feed forward compensation to the TL431 and a small pole at the feedback pin of the control IC. The actual compensation values will need to be determined by taking the prototype supply into the laboratory and making measurements and adjustments.

The selection of current sense components and under-voltage components are the same as the MAX5052 example in the flyback converter section.

Non-Isolated Forward Converter Example

The current levels in the flyback example for automotive use were quite high. The input current consists of very large, short pulses. The output also consists of very large, short pulses. A forward converter can reduce both the output ripple and the input ripple by allowing the duty cycle to be larger than 50%. Our next example, Figure 5-17, shows how to implement such a supply.

The duty cycle in an off-line forward converter is limited to 50% by the voltage required to reset the flux in the transformer and the switch breakdown voltage. At 50% duty cycle, the reverse voltage can be equal to the input voltage. In our automotive application, we can use a high voltage switch to advantage. The high reverse voltage will allow the flux in the transformer to reset in a very short period of time.

We start from the same set of requirements as the flyback example and use the same control IC. We choose the same 167 kHz operating frequency for a cycle time of 6 μs.

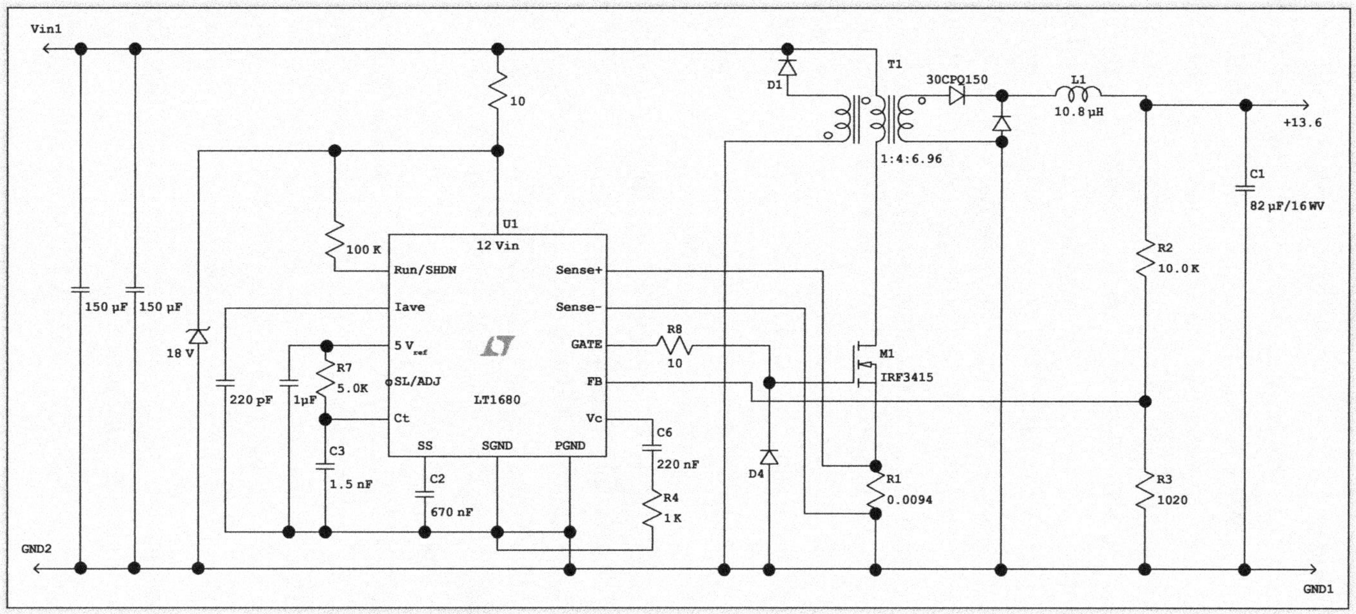

Figure 5-17: Non-isolated forward converter with 13.6V output

We can set the maximum duty cycle to 75% at 11.0 V input. The data sheet shows that the maximum duty cycle will vary from IC to IC, from about 70% to about 78%, when we set the nominal value to 75%. Our calculations will need to allow for 80% duty cycle as the worst case. The volt-seconds during switch on-time will need to equal the volt-seconds when the switch is off. The ratio of on-time to off-time is 80/20, so the reverse voltage on the transformer primary during off-time will be four times the input voltage. This sets the turns ratio for the clamp winding to 4:1. The switch withstand voltage will be five times the input voltage (4× for the clamp plus 1× for the input supply) at the highest input voltage. This gives a minimum value of 15.0 V * 5 = 75 V. A check of the International Rectifier website shows either 100 V or 150 V MOSFETs. It probably makes the most sense to choose a 150 V device to ensure margin in the presence of transients. The IRF3415 is a TO-220 package that has 150 V V_{DSS}, 42 mΩ on resistance, and 43 A I_{DSS}. The IRF3315 is a similar and less expensive part, but it only has 15 A I_{DSS} at 100°C.

A 150 V Schottky diode is a reasonable first try at the output rectifier. The turns ratio of the transformer is likely to be very close to 1.5:1 for primary to secondary, since our goal is to reduce the input ripple and output ripple. However, we are allowing the reverse voltage during transformer reset to be four times the input voltage. This means the reverse voltage on the diodes will be four times the input times the turns ratio. This will require a diode with at least 90 V PRV. The 150 V PRV rating will allow margin for a transformer ratio up to 2.25:1.

We can use the IRF 30CPQ160 150 V PRV/30 A diode. This is the same diode family that we used in the flyback example. We can choose the peak output current as 11 A with 2 A of ripple current. Each diode will pass part of the total current, so the forward voltage drop will be 0.75 V for the whole cycle. The maximum output power will be 13.6 V * 10.0 A + 0.75 V * 10.0 A = 143.5 W.

We use a rearranged version of the buck converter equation from Chapter 1 to determine the required input voltage.

$$V_{IN} = (V_{OUT} + V_{Diode})/DC = 14.35 \text{ V}/0.75 = 19.1 \text{ V}$$

This is the voltage that must be present at the secondary winding at the lowest input voltage. This gives the turns ratio of the transformer:

$$N = 19.1 \text{ V}/11.0 \text{ V} = 1.74$$

We can verify the required duty cycle at high input voltage. The input voltage will be:

$$15 \text{ V} * 1.74 = 26.1 \text{ V}$$

This means the duty cycle at high input will be 14.4/26.1 = 55%. The supply will require slope compensation over the entire operating range. The high input voltage of 15.0 V times the (4:1 * 1.74:1) turns ratio yields 104 V reverse voltage. This confirms that the diode we chose is adequate.

The inductor value is determined by the ripple current, applied voltage, and duty cycle. The applied voltage is the transformer voltage less the diode drop. We apply the inductor equation:

$$L = V \frac{dt}{dI} = (18.4 - 13.6) * \frac{0.75 \times 6 \text{ } \mu s}{2.0 \text{ A}} = 10.8 \text{ } \mu H$$

The output capacitor value is determined by the ripple voltage requirement. We have 300 mV of ripple and 2.0 A of ripple current. We can choose ESR and the capacitor value using our one-third and two-thirds rule:

$$ESR = \frac{200 \text{ mV}}{2.0 \text{ A}} = 100 \text{ } m\Omega$$

The target capacitance is:

$$X_C = \frac{100 \text{ mV}}{2.0 \text{ A}} = 50 \text{ } m\Omega$$

$$C = \frac{1}{2 * \pi * 167 \text{ kHz} * 50 \text{ } m\Omega} = 19 \text{ } \mu F$$

The ripple current and ESR requirements are easily met with a single 82 μF/16 WV Panasonic WA series polymer electrolytic capacitor. This capacitor has 39 $m\Omega$ ESR and 2.5 A ripple current rating in a surface mount package. The RMS ripple current is approximately equal to one-half the P–P ripple for a triangular wave, so our output ripple current is approximately 1 A.

The average input current is 141 W/11.0 V = 12.8 A. The input current is essentially a rectangular pulse of 12.8 A/0.75 = 17 A. The RMS current is

$$I_{RMS} = I_{IN} (DC - DC^2)^{1/2} = 12.8 (0.75 - 0.56)^{1/2} = 5.6 \text{ A}$$

Two of the 150 μF/20 WV Panasonic WA series polymer electrolytic capacitors will do nicely for the input filter. This capacitor has 26 mΩ ESR and 3.7 A ripple current rating in a surface mount package. This is quite a contrast to the ripple requirements of the flyback design, where the input RMS ripple was 9 A RMS and the output ripple was 4.8 A RMS. We require fewer and less expensive filter capacitors by changing from a flyback circuit to a forward converter.

The current sense resistor is set by average current rather than peak current for this control IC. The equation is found in the data sheet:

$$R_{CS} = 120 \text{ mV/}I_{AVG} = 0.12 \text{ V/}12.8 \text{ A} = 9.4 \text{ m}\Omega$$

The average current limit is set by the combination of the current sense resistor and the current limit integration capacitor. The data sheet recommends setting this capacitor to 220 pF.

The output voltage and soft start calculations are the same as for the flyback example.

We want to restrict the maximum duty cycle to 75%, so we choose the 5 K timing resistor from the graph in the data sheet. Another graph on the data sheet indicates that 1.5 nF will yield 167 kHz operation for this timing resistor. A duty cycle above 50% requires slope compensation for all current mode controllers. The LT1680 provides internal slope compensation that should be adequate for our example supply.

Push-Pull Circuits

Push-pull circuits are not well suited to voltage mode IC controllers because any flux imbalance in one leg of the transformer primary will eventually saturate the transformer core. A current mode controller will control the imbalance and limit the current through both legs. One switch and one winding of the transformer may still carry more load than the other, but the total flux in the core is controlled by limiting the maximum current in each winding.

Figure 5-18 shows a representative push-pull converter. Notice that the secondary side uses a center-tapped full wave rectifier configuration. A push-pull circuit requires full wave rectification. Most practical circuits use a center-tapped

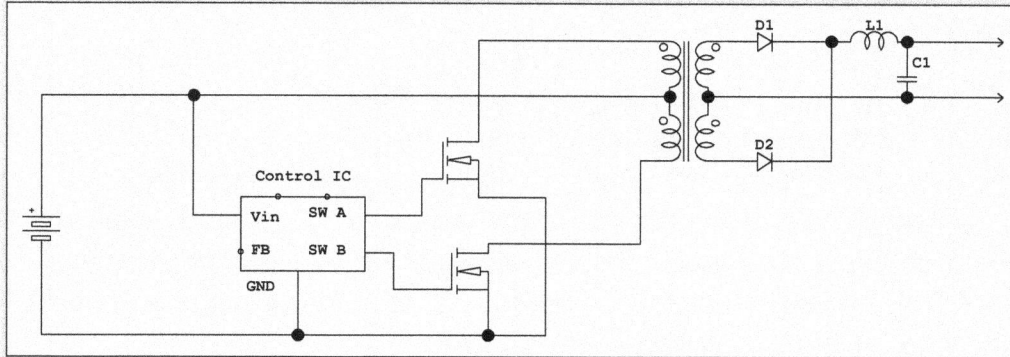

Figure 5-18: Representative push-pull converter

transformer and a dual diode, since there is only one diode drop during each half-cycle and two diodes. It is possible to use a full wave bridge to simplify the transformer, but then the voltage drop for each half-cycle is two diode drops and it uses four diodes. Basically, copper is much cheaper than silicon.

Push-pull circuits fell out of favor when the only IC controllers were voltage mode because of the problems with transformer balance. They are more popular for moderate power circuits now that current mode controllers are readily available. Push-pull circuits are popular at all power levels for point of load applications where the voltage stress on the switch is not an issue.

The primary requires twice as many turns as a bridge circuit, so the transformer is more complicated than a half bridge transformer. The switches must withstand twice the input voltage where the switch voltage for a half bridge is equal to the input voltage. The biggest advantage of push-pull over half bridge is that neither switch requires isolated drive. A clamp circuit is not necessary in a push-pull circuit because one of the output diodes will continue to conduct when both switches are off. This allows magnetizing inductor current to flow while current in the output choke ramps down. The magnetizing inductance current will be forced to zero when the alternate switch closes.

The effective switching frequency is double the oscillator frequency. Each switch provides the equivalent of a single switch forward converter. The bipolar drive doubles the effective duty cycle and the operating frequency of the output filter is double the switching frequency.

The control IC must provide two-phase output pulses to alternately drive the switches. Additionally, the circuit will behave badly if both switches conduct at the same time. The transformer will allow very large switch currents to flow if both switches conduct at the same time. A push-pull control IC must provide the ability to set a proper amount of dead time between the alternate phases. This will ensure that one switch is off before the other switch begins conducting.

Practical Push-Pull Circuit Design

Typical steps for designing a push-pull converter are listed below:

1. Choose a controller IC based on power level and bill of material constraints.

2. Choose the switching frequency.

3. Use the input voltage range goal to select the maximum duty cycle goal.

4. Pick the output diodes.

5. Calculate the output inductor value.

6. Design the transformer winding ratios.

7. Determine the maximum power and pick the switches.

8. Choose the output capacitor based on ripple requirements.

9. Design the auxiliary supply, if needed.

10. Design the ancillary IC components, including the feedback circuit.

Our push-pull example is a telecom supply that converts 48 V to an isolated 5 V/20 A supply with 100 mV of ripple. Figure 5-19 shows the circuit we are designing.

A search for control ICs designed specifically for push-pull or bridge operation yields very few parts. Most of the first- and second-generation current mode controllers (such as the 1846) provide the necessary functions, but they need a large number of external components for a working supply. There are not many

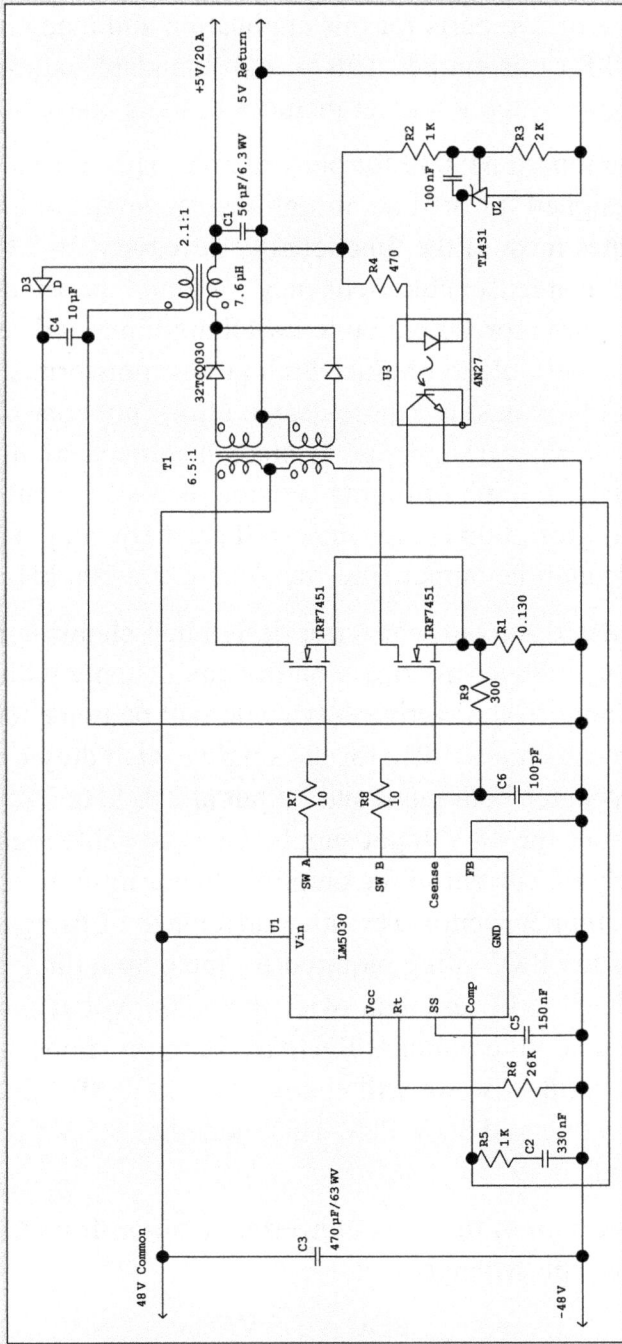

Figure 5-19: An isolated push-pull design

modern control ICs designed for push-pull and bridge operation. Some manufacturers have only one or two parts for this application and many have no modern products at all for this market. This is understandable, since only a very small portion of the power supply market includes designs above 200 W.

We will choose the National LM5030 for our example. This part is a 10-pin surface mount part designed for off-line or high voltage applications. Like most modern ICs, it integrates most of the functionality necessary for a low part count design. 200 kHz is a reasonable frequency for a high power system. Higher frequencies require more attention to switching times and second- and third-order effects. It is possible to produce high power transformer designs at higher frequencies, but you must put more design effort into controlling elements like layout, transformer design, and semiconductor selection. The LM5030 data sheet gives a graph of timing resistor versus frequency. The graph shows 26 K will give 200 kHz for the oscillator frequency. The switches will switch at 100 kHz and the output filter will operate at 200 kHz.

We saw in the non-isolated forward converter design that choosing a duty cycle above 50% significantly reduces the ripple on the input supply without any effect on the output ripple. The effective duty cycle will be twice the single switch duty cycle. We can choose 40% for the single switch duty cycle. This yields an 80% duty cycle for both input and output at 200 kHz. This large duty cycle is adequate because the 48 V input supply has reasonable regulation. We will choose 1.0 A of ripple current in the output inductor to minimize the ESR requirements of the output capacitor. The other advantage of using a large duty cycle is that it reduces the PRV rating required for the output diodes. We can expect the reverse voltage to be less than twice the output voltage, so 20 V PRV diodes should be adequate. International Rectifier Schottky diodes are available in either 15 V or 30 V ratings, so we will choose the 32TCQ030 dual diode. This device has 30 A rating and 30 V PRV. This diode has 0.5 V forward drop at 20 A forward current.

We use a rearranged version of the buck converter equation from Chapter 1 to determine the required input voltage.

$$V_{IN} = (V_{OUT} + V_{Diode})/DC = 5.5 \text{ V}/0.80 = 6.9 \text{ V}$$

This is the voltage that must be present at the secondary winding at the lowest input voltage. This gives the turns ratio of the transformer:

$$N = 48 \text{ V}/6.9 \text{ V} = 6.96$$

It will simplify the design significantly if we adjust this to 6.5 and use a slightly smaller duty cycle. This value will make the transformer design reasonable. We will probably design the transformer with either two or three turns for each leg of the secondary, which will require either 13 or 20 turns per primary winding.

The inductor value is determined by the ripple current, applied voltage, and duty cycle. We apply the inductor equation:

$$L = V\frac{dt}{dI} = (6.9 - 5.0) * \frac{0.8 \times 5 \text{ μs}}{1.0 \text{ A}} = 7.6 \text{ μH}$$

The output capacitor value is determined by the ripple voltage requirement. We have 100 mV of ripple and 1.0 A of ripple current. We can choose ESR and the capacitor value using our one-third and two-thirds rule:

$$ESR = \frac{67 \text{ mV}}{1.0 \text{ A}} = 67 \text{ mΩ}.$$

The target capacitance is:

$$X_C = \frac{33 \text{ mV}}{1.0 \text{ A}} = 33 \text{ mΩ}.$$

$$C = \frac{1}{2 * \pi * 200 \text{ kHz} * 33 \text{ mΩ}} = 24 \text{ μF}$$

A 56 μF/6.3 WV Panasonic S series surface mount polymer electrolytic capacitor has only 9 mΩ ESR, so the ripple will be significantly below the 100 mV target. This capacitor has 3 A ripple current rating.

The input supply power is approximately 117 W (85% efficiency). The average input current is 2.4 A and the peak input current is 3.1 A with an 80% duty cycle. The RMS current is 1.24 A. A Panasonic FC series 470 μF/63 WV capacitor will provide low ESR (under 1 Ω) with adequate ripple current rating. The switches will need to handle at least twice the input voltage. The closest V_{DSS} rating with margin is 150 V. The IRF3415S is a D2PAK device that has more than enough current rating and adequate voltage rating. The IRF7451 is a device in an SO-8 package. It has a continuous drain current rating of 3.6 A.

Since the average current for each switch is one-half the total, this device might be adequate if small size is a design goal.

The feedback circuit uses a TL431 to drive a 4N27 optoisolator. There is loop compensation on both the TL431 and on the compensation pin of the control IC. The control IC implements internal slope compensation so no external slope compensation should be necessary.

The V_{CC} supply is different from any we have used so far. We put an auxiliary winding on the main filter inductor to use it the same way we used the inductor in the flyback supplies. Note that the auxiliary supply winding is polarized so that the supply charges while the current is discharging in the filter choke. The voltage across the inductor will vary depending on the 48 V input voltage while the filter choke is charging with current. However, when the filter choke is discharging, the diodes clamp the voltage across the choke to approximately the output voltage. Since the output voltage is highly regulated, we get a well regulated auxiliary supply for the control IC. This would seem to be the best possible source for IC power. The problem, especially for off-line supplies, is that the safety isolation between the windings of the filter inductor must be the same as the safety isolation of the main transformer. The supply voltage of our example will be (2.1 * 5 V – 0.7 V) or 9.8 V. The IC supply voltage will change only slightly with changes in output current level.

The current sense resistor is calculated using the information in the data sheet:

$$R = 0.5/I_{PK} = 0.5/3.8 = 0.130 \ \Omega.$$

We implement a small RC filter at the current sense pin in order to eliminate false setting of the current sense because of transients.

The soft start pin provides a current source of 10 μA. This current source charges the soft start capacitor to 0.5 V. If we use 30 ms for soft start, we need a 0.15 μF soft start capacitor.

Half Bridge Circuits

Half bridge circuits are the topology of choice for off-line converters between 200 W and 1000 W. Figure 5-20 shows a representative half bridge converter.

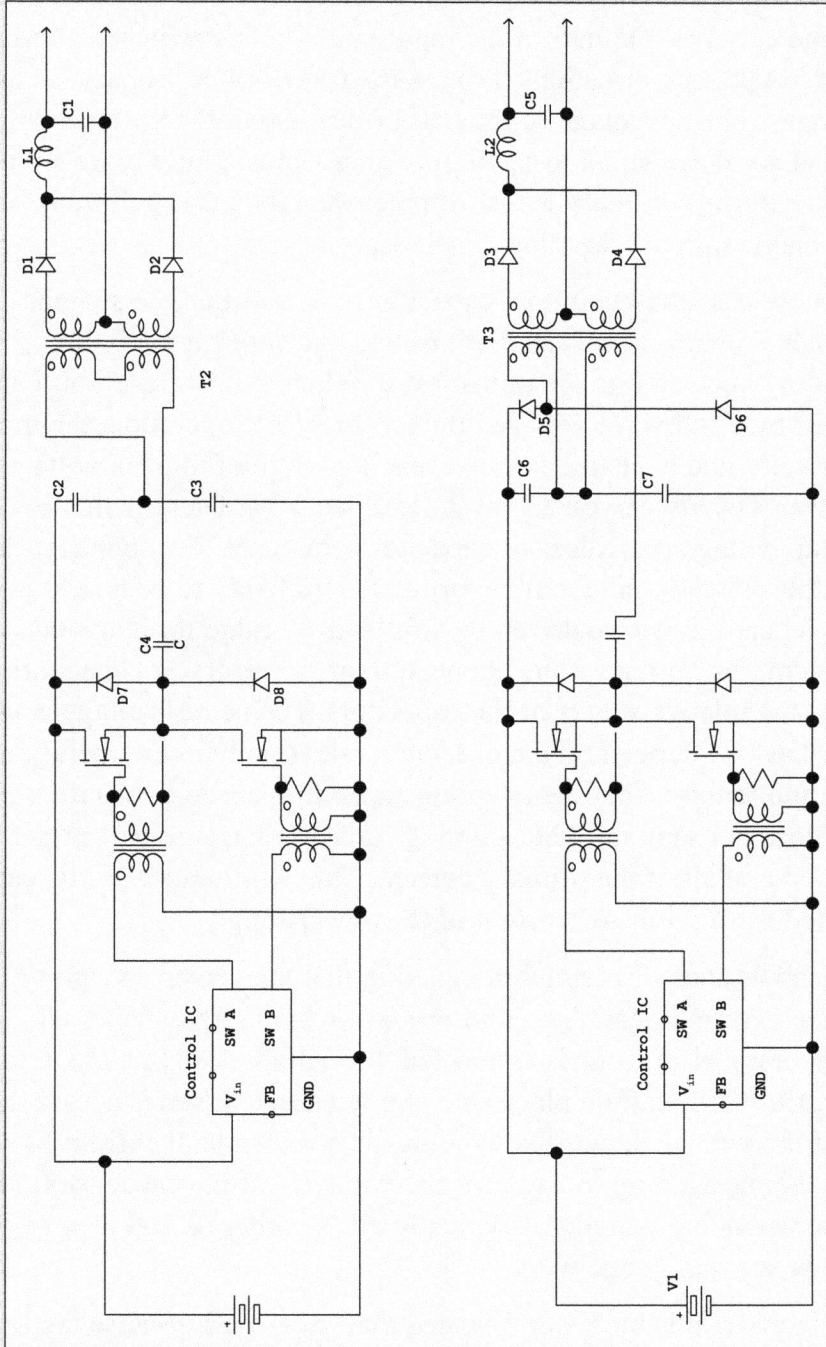

Figure 5-20: (a) Representative half bridge circuit for full wave doubler operation. (b) Equalizing circuit for use with full wave bridge input

159

The capacitive voltage divider (C2, C3) is an integral part of the circuit. It provides a voltage equal to one-half of the input voltage. The switches alternately drive current in opposite directions through the transformer primary, as in the push-pull circuit. The advantage of the half bridge is that the switches must only withstand a voltage equal to the input voltage plus a little more for transients. The transformer primary is also simpler than the push-pull transformer, since only a single primary winding is necessary.

Notice that there is a small coupling capacitor (C4) between the switches and the transformer primary. This capacitor ensures that flux cannot build up in the primary winding and saturate the transformer. When the two reservoir capacitors are driven by a full wave voltage doubler for 115 V operation, the input supply diodes alternately charge up the capacitors to the full peak voltage of the input power. The voltage on each capacitor has a hard supply that ensures a hard center tap voltage regardless of capacitor symmetry. The coupling capacitor between the switches and the transformer is less likely to be necessary. However, if the capacitors are driven by a full wave bridge for a universal input or 240 V system, the voltage at the connection of the reservoir capacitors will be a factor of the relative values of the capacitors. The center voltage is now a "soft" value that will depend on the capacitor values and circuit operation. "Soft" operation requires that the coupling capacitor is used to ensure that the transformer does not saturate. The coupling capacitor has one-half of the input voltage applied and all of the primary current. This will require an AC capacitor that is rated for the full AC current of the power supply.

Figure 5-20 shows another method of ensuring that the center voltage of the capacitors stays symmetrical. A second primary winding (balancing winding) with the same number of turns is connected through diodes D5 and D6 to the input supply. Circuit operation places the two windings in series across the two capacitors. If the voltage across the windings is not identical, current flows from the balancing winding to equalize the voltages on the capacitors. The current in the balancing winding is typically on the order of 100 mA, so the winding can be a small gauge wire.

The half bridge circuit is more complicated than push-pull because the top switch requires isolated drive. Current mode control requires that a current transformer be placed in series with the primary winding. The current sense also requires full wave rectification to sense the current of each switch. Notice that there are clamp diodes across each of the switches. It is possible to use the body-drain diode of the MOSFETs, but these diodes have poor turn-on and turn-off characteristics. It is good practice to use high speed diodes to prevent the MOSFET diodes from conducting.

Practical Half Bridge Circuit Design

The design of half bridge and full bridge circuits contains the same steps. Typical steps for designing a bridge converter are listed below:

1. Choose a controller IC based on power level and bill of material constraints.

2. Choose the switching frequency.

3. Use the input voltage range goal to select the maximum duty cycle goal.

4. Pick the output diodes.

5. Calculate the output inductor value.

6. Design the transformer winding ratios.

7. Determine the maximum power and pick the switches.

8. Choose the output capacitor based on ripple requirements.

9. Design the auxiliary supply, if needed.

10. Design the ancillary IC components, including the feedback circuit.

Our half bridge example is a 12.0 V/40 A off-line universal power supply. The ripple goal is 100 mV. Figure 5-21 shows our example power supply. The National LM5030 is a good candidate for our supply. Once again, we choose 100-kHz operation to simplify design but still give good efficiency.

We must start our design at 100 VDC input to design the inductor for maximum duty cycle. We will choose a maximum duty cycle of 40%. We also set our ripple current goal at 4.0 A. The output diode reverse voltage at 100 V input is likely to be 18 V. The reverse voltage at 390 V input will then be approximately 70 V. The International Rectifier 80CNQ080A has 80 A current

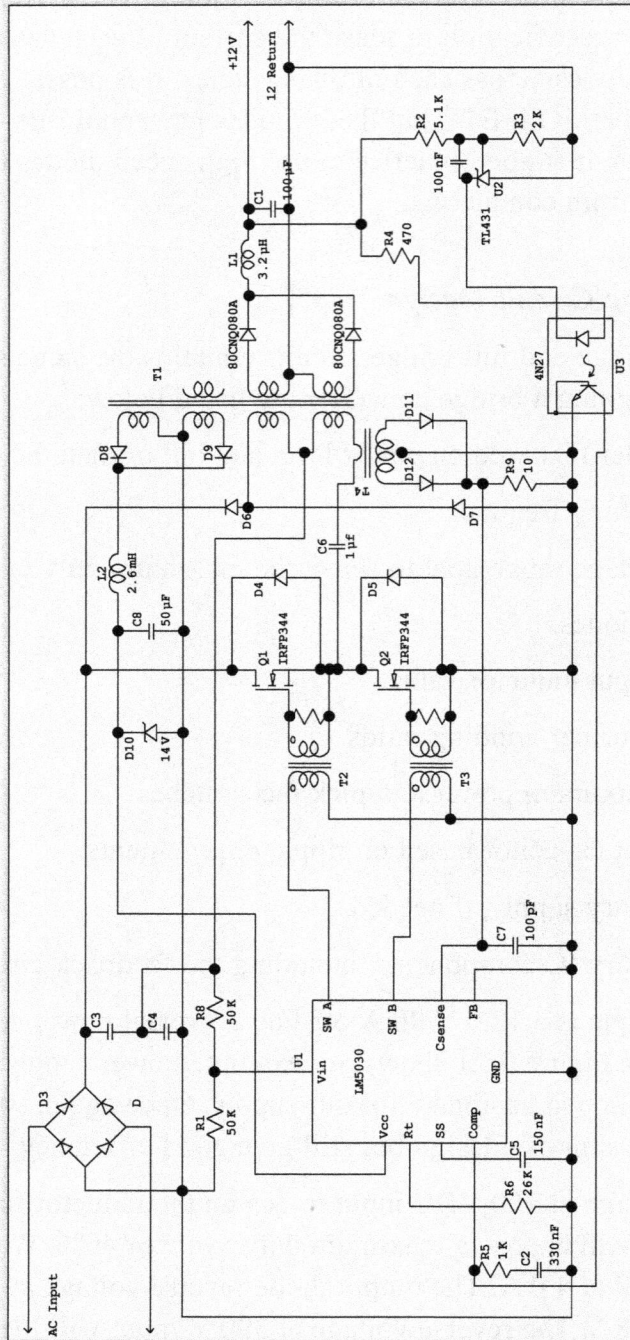

Figure 5-21: Half bridge design 12.0 V/40 A off-line universal power supply

rating and 80 V PRV. The forward voltage at 40 A is 0.8 V, so this diode will dissipate 32 W at full output.

We use a rearranged version of the buck converter equation from Chapter 1 to determine the required input voltage:

$$V_{IN} = (V_{OUT} + V_{Diode})/DC = 12.8 \text{ V}/0.80 = 16.0 \text{ V}.$$

The maximum voltage at the rectifiers will be 62.4 V, so our diode selection is adequate. The duty cycle at high input voltage will be 20% in the output circuit at high input voltage.

The output inductor will be:

$$L = V\frac{dt}{dI} = (15.2 - 12.0) * \frac{0.8 \times 5 \text{ μs}}{4.0 \text{ A}} = 3.2 \text{ μH}.$$

Remember that the voltage across the transformer is only one-half of the input voltage. This gives the turns ratio of the transformer:

$$N = 50 \text{ V}/16.0 \text{ V} = 3.2.$$

The losses in the circuit are quite large. The diode losses are the main contribution to power loss in the switching circuit. We should add at least 20 W more to account for other losses in the circuit. This gives a total switching power input of 532 W. The input current at low input voltage will be 5.32 A average or 6.65 A peak. The switches will need 450 V V_{DSS} and at least 7 A I_{DSS}. The IRFP344 has 450 V V_{DSS} and 9 A I_{DSS} with 0.63 Ω on resistance. The total gate charge is 60 nC for this switch, so gate current will be 12 mA.

The output capacitor value is determined by the ripple voltage requirement. We have 100 mV of ripple and 1.0 A of ripple current. We can choose ESR and the capacitor value using our one-third and two-thirds rule:

$$ESR = \frac{67 \text{ mV}}{4.0 \text{ A}} = 17 \text{ mΩ}.$$

The target capacitance is:

$$X_C = \frac{33 \text{ mV}}{4.0 \text{ A}} = 8.3 \text{ mΩ}.$$

$$C = \frac{1}{2 * \pi * 200 \text{ kHz} * 8.3 \text{ mΩ}} = 96 \text{ μF}$$

Once again, we see that the filter capacitor is quite small, even for a very large output current.

We will need an auxiliary supply for this design. The output of the IC supply will have no relationship to the output voltage. The best we can do is to design a circuit that is close to the required voltage and regulate it to the required voltage. The combination of the IC current and the switch current is only 15 mA. We can design the supply to deliver 12 V and use a zener diode to ensure that the voltage does not rise above the 16 V maximum of the IC. The winding will have the same number of turns as the main output, but we can use smaller wire that produces a convenient winding. We set the ripple current to 5 mA.

$$L = V\frac{dt}{dI} = (15.2 - 12.0) * \frac{0.8 \times 5 \ \mu s}{5 \ mA} = 2.6 \ mH$$

Any convenient switchmode electrolytic capacitor with 50 μF will have sufficiently low ESR and produce a low ripple supply for the IC. We can choose a resistor to charge the IC supply to the required 7.7 V to start operation. The IC stops operating if the IC supply voltage falls below 6.1 V, so we will need a large capacitor to supply current until the bootstrap charges the capacitor. The current will be quite large at startup, so it should only take two or three cycles to begin supplying the necessary current. A resistor to supply 1 mA of current should provide adequate start time at low input voltage.

The current sense for this circuit is significantly different from what we have seen so far. All of our examples have used a sense resistor referenced to ground. Bridge circuits need a current sense transformer (T4) connected in series with the transformer primary. It is also possible to measure the output inductor current directly using a current transformer, but the current sense transformer would then require full safety isolation certification. The current sense transformer uses full wave rectification to allow measurement of the current from both switches.

The voltage feedback circuit is the same one we used for the push-pull example.

Full Bridge Circuits

Full bridge circuits are useful for power supplies that operate above 500 W. They are the most complicated of all off-line supplies and, therefore, the most

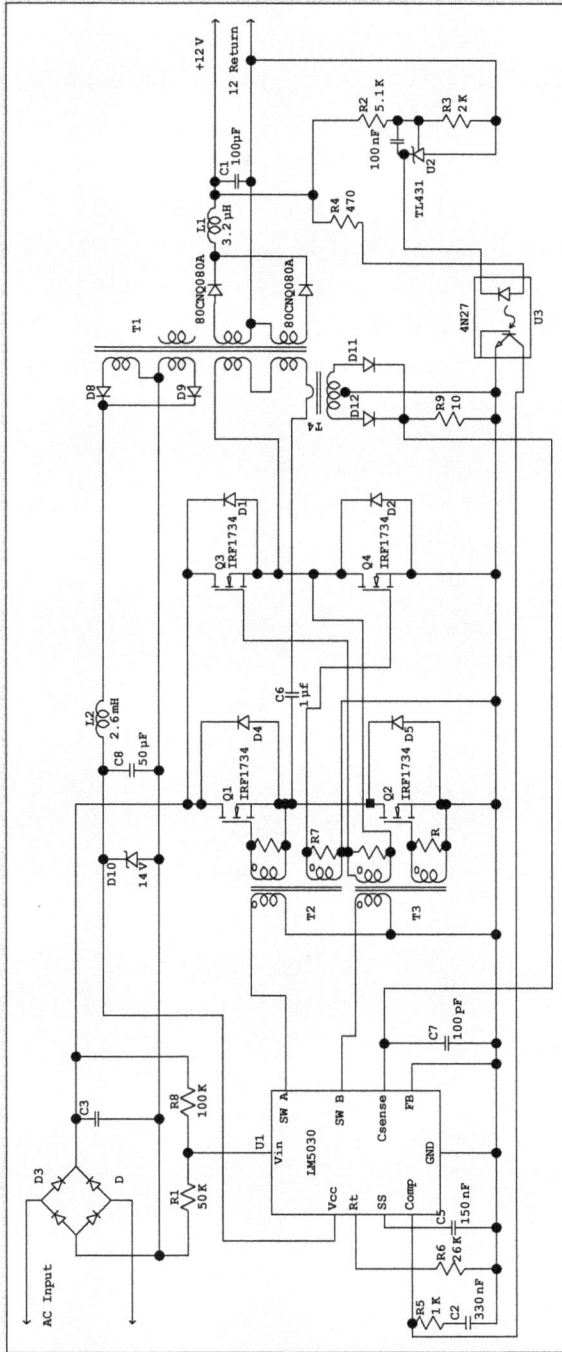

Figure 5-22: Full bridge design (transformation of Figure 5-21)

expensive. Full bridge operation is only chosen when the primary current is too large for two switches to handle. A full bridge circuit replaces the two capacitors with two switches and clamp diodes. Both top switches will require isolated drive. If the transformer turns calculations, the full input voltage will be used, rather than one-half, as in the half bridge. The full bridge design uses one power line capacitor instead of two. The capacitance of the single capacitor (C3) is smaller than in the half bridge. The reduced cost of one smaller capacitor versus two large ones offsets the extra cost of the semiconductors for full bridge operation. Additionally, the switches can be less expensive because the current level is one-half that needed for a half bridge. Figure 5-22 shows our example full bridge converter.

We can redesign the half bridge example above to be a full bridge circuit. The main design decisions and calculations remain the same for this new example. The first change will be the transformer turns ratio.

$$N = 100 \text{ V}/16.0 \text{ V} = 6.3$$

The next change will be switch selection. The input current will be 2.66 A or 3.33 A peak. This lower current allows us to choose a cheaper switch. The IRFP344 is $2.33 each (100's) versus the IRF1734, which is $0.94 each (100's) in 2004 dollars. The IRF1734 is a 450 V/3.4 A switch that will have enough margin in this application. If the transformers in Figures 5-21 and 5-22 use identical windings for the two primary windings, the same transformer can be used in both circuits.

The schematics for both of the bridge circuits are extremely complicated compared to the single switch circuits and the push-pull circuit. The pulse transformers must drive two transistors each, where the half bridge transformer only drives one transistor.

CHAPTER 6

Passive Component Selection

- Capacitor Characteristics
- Aluminum Electrolytic Capacitors
- Solid Tantalum and Niobium Capacitors
- Solid Polymer Electrolytic Capacitors
- Multilayer Ceramic Capacitors
- Film Capacitors
- Resistor Characteristics
- Carbon Composition Resistors
- Film Resistors
- Wire Resistors

Passive Component Selection

A capacitor is a capacitor is a capacitor if you are building a low frequency or low power analog circuit. This is not true for switching supplies. The high currents and high frequencies have significant consequences for the capacitors we choose. We have already looked at capacitors in Chapter 3, when we discussed aluminum electrolytics for input supplies and capacitors for EMI filtering. Here we will look at the details of various types of capacitors that are suitable for switchmode power supplies.

Capacitor Characteristics

The main characteristic of a capacitor is that it stores charge when presented with a voltage. However, we can model a real capacitor as a combination of resistance, capacitance, and inductance. Figure 6-1 shows equivalent circuits for a real polar and nonpolar capacitor.

Note that the polar capacitors also include a parasitic diode that allows current to flow if the capacitor is reverse biased. This diode is an actual physical diode formed by the metal and oxides used for the plates and the dielectric. The polar capacitors are formed from metals that are called "valve metals" because of the characteristics of the metal and the oxide dielectric. The valve metals used for capacitors are aluminum, tantalum, and niobium.

The primary effects we must consider for circuit performance are the capacitance, equivalent inductance (ESL), and equivalent resistance (ESR). We must also consider failure modes for each capacitor to ensure that the system has adequate reliability. Each capacitor type has a different set of failure modes.

Dissipation factor, tan δ, and impedance are the three parameters that you will find in catalogs and data sheets to describe the loss characteristics of capacitors.

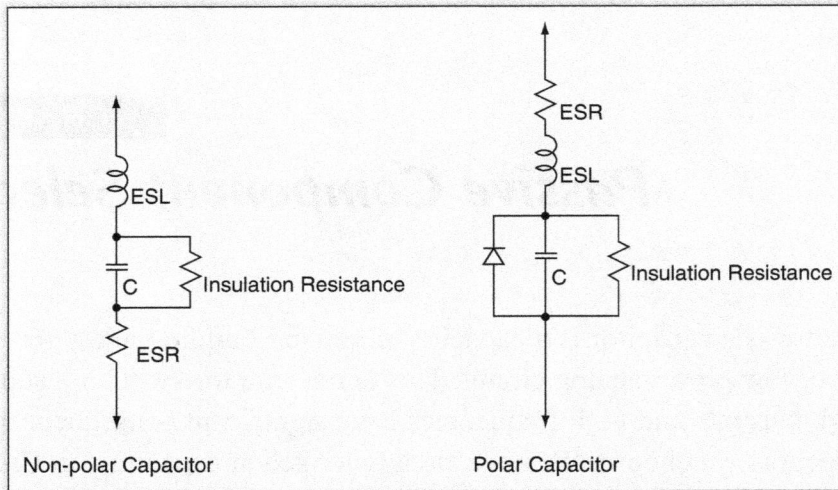

Figure 6-1: Equivalent circuits for a real polar and nonpolar capacitor

Figure 6-2 shows the relationships of the values that determine the loss parameters. Notice that all three parameters are dependent on X_C and X_L. Since X_C and X_L are frequency-dependent, the loss parameters are also frequency-dependent. Also, note that any external factor, such as applied voltage or temperature, that affects capacitance will affect the loss parameters. Dissipation factor is expressed in percent:

$$DF = \frac{ESR}{X_C} \times 100$$

Our rule of thumb for ripple voltage requires a DF of 67% or larger.

Tan δ and power factor are mirrors of each other. Power factor is the cosine of the angle formed by the magnitude of impedance and the ESR. Tan δ is the tangent of the adjacent angle.

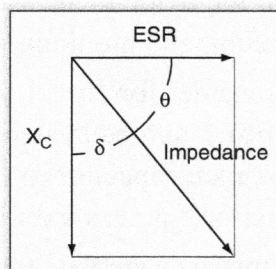

Figure 6-2: Relationships of the values that determine the loss parameters of capacitors

$$\tan\delta = \left(\frac{\text{ESR}}{\text{X}_\text{C}}\right)$$

Impedance is the length of the vector formed by ESR and X_C in combination with X_L. All of our examples in Chapters 4 and 5 used ESR and X_C to determine ripple voltage. We could just as easily have done a single calculation using magnitude of impedance to approximate the ripple voltage.

Ripple current capability is related to ESR. Real power is dissipated in ESR due to AC current flowing through the capacitor. The power dissipation causes increased temperature in the capacitor. Each capacitor technology has different capabilities for power dissipation and temperature rise.

Aluminum Electrolytic Capacitors

Aluminum electrolytic capacitors are the oldest bulk capacitor technology. They are the composition of choice for the input power supply in an off-line power supply. Electrolytic capacitors can be manufactured with large capacitance, large voltage rating, and small size for power line and audio frequency operation. These three characteristics are the primary advantages of aluminum electrolytics.

All aluminum electrolytic capacitors are made of a sandwich of an aluminum anode foil, a paper separator, an aluminum cathode foil, and another paper layer. This sandwich is rolled into a coil and held in a sealed container. Wires are welded to the anode and cathode foils to make a connection to the external circuit. The cathode and anode foils are chemically etched to increase the surface area of the foil in order to increase capacitance. The increase in capacitance is about 20 times for high voltage capacitors and as much as 100 times for low voltage capacitors. The dielectric of an aluminum electrolytic is composed of aluminum oxide formed on the surface of the anode foil. The depth of the oxidation controls the voltage rating and the capacitance of the final assembly. The oxide layer is formed by pulling the foil through an electrolyte bath with voltage applied between the bath and the foil. The cathode foil is etched to expose more aluminum to allow better electrical contact with the liquid electrolyte. The liquid electrolyte is actually the negative terminal of the capacitor.

The sandwich is rolled into a coil, which is then immersed in an electrolyte solution of solvent and salts. The solvent is typically ethylene glycol, dimethyformaldehyde, or gammabutyrolacetone, and the salt is typically ammonium borate or other ammonium salts. The solvent used determines the temperature rating of the capacitor. After the coil absorbs the electrolyte, it is placed inside an aluminum can. Small capacitors are manufactured with a rubber plug at the bottom of the can to seal in the electrolyte and to provide a safety vent. Larger capacitors have a phenolic or nylon cover with an O-ring to seal the can. The safety vent is usually part of the plastic seal. The safety vent can also be created by scoring the aluminum of the can so that the can will fracture along the score lines to create a vent. The electrical connections are made through the bottom seal of the capacitor. A small amount of water in the electrolyte allows the capacitor to self-heal. If a fault occurs, the current will break the water into hydrogen and oxygen. The oxygen reacts with the aluminum to form new aluminum oxide and repairs the capacitor. The hydrogen vents to the atmosphere.

Heat is the primary source of failure in aluminum electrolytic capacitors. Once the core of the capacitor reaches the boiling point of the electrolyte, the internal pressure will rise and the vent will allow some of the electrolyte to escape. The loss of electrolyte causes the ESR to increase, which then causes the capacitor to dissipate more heat. This positive feedback can cause rapid failure of a capacitor at high temperatures. Electrolytic capacitors fail open because of loss of electrolyte.

The diode in the equivalent circuit is a zener diode. As the voltage increases beyond the rated voltage, the capacitor will eventually begin conducting current and the voltage will remain relatively constant. The combination of current and voltage will result in a temperature rise that will result in failure. The reverse voltage of the equivalent diode is approximately 1.5 V and will likewise result in failure due to temperature rise if the capacitor is reverse biased.

The dielectric constant decreases with frequency, so capacitance also decreases with increasing frequency. The electrolyte is also a primary contributor to the ESR of the capacitor. The ESR decreases with increase of both frequency and temperature.

It is very important to factor in the temperature coefficient of ESR and capacitance for low temperatures. The ESR typically increases 100-fold at −40 C°. The capacitance value can decrease by as much as 40% at −40 C°, depending on the temperature rating of the capacitor. The increase in ESR will also decrease the ripple current rating of the capacitor at low temperatures.

The primary reliability concern for both switchmode filter capacitors and off-line input filter capacitors is ripple current rating. The ripple current through the ESR creates heat that increases the core temperature of the capacitor. A capacitor that is operated near its temperature limit may only last a week or two before failure. It is important when selecting an aluminum electrolytic to use the derating and life figures on the manufacturer's data sheet to ensure that the supply will meet the reliability requirements. It is common for a capacitor rated at 85 C° to only be rated for 2000-hour life at 85 C°. Such a capacitor would need significant temperature derating to have a useful life of several years in continuous service. Most manufacturers produce lines of capacitors suitable for elevated temperature service. It is typical to use a capacitor that is rated for 2000-hour life at 105 C° in an application with maximum temperature of 80 C°. Running the capacitor 25 C° below the rating would yield a life on the order of 1000 times the life at rated temperature.

Aluminum electrolytic capacitors are usually not surface mount components. The reflow temperatures will boil the electrolyte and the chemicals for vapor phase will etch the outer can. Some manufacturers make surface mount electrolytic capacitors, but the process controls for automatic soldering are very rigid.

The inductance of an aluminum electrolytic is due primarily to the inductance of the lead wires. SMT devices have the lowest ESL (on the order of 20 nH) and the axial lead types will have the greatest values (on the order of 200 nH). ESL is typically not an issue because the ESR dominates the magnitude of the impedance at switching frequencies.

Solid Tantalum and Niobium Capacitors

Solid tantalum capacitors are used in applications where their volumetric efficiency is an advantage. Tantalum capacitors have significantly lower maximum voltage than aluminum electrolytics. The maximum working voltage varies from 30–50 V, depending on the manufacturing process. Niobium capacitors are manufactured in a manner similar to solid tantalum capacitors and have similar characteristics. A major advantage for surface mount tantalum and niobium capacitors is that they will survive reflow temperatures.

Solid tantalum capacitors use metallic oxide as the dielectric, in a manner similar to aluminum electrolytics. The anode of a tantalum capacitor is initially formed from a mixture of a binder and small particles of tantalum metal. The mixture is pressed into a slug with a tantalum wire embedded in it. Then the slug is heated to drive out the binder, leaving a porous metallic structure with a very large surface area. The slug is sintered at high temperature to fuse the particles of tantalum into a porous solid structure. The dielectric of tantalum pentoxide is formed by immersing the slug in an acid bath and passing current through the slug and the bath at high temperature. Current and time control the thickness of the oxide layer and control the capacitance created. Impurities in the surface oxide cause leakage current when the capacitor is in use. The oxide formation step also creates a layer of tantalum oxide between the tantalum metal and the pentoxide layer. This structure creates a metal-insulator-semiconductor diode structure that is similar to the structure of a Schottky diode. It is this structure that makes the capacitor an actual physical diode when reverse biased. The slug is dipped in a bath of manganese nitrate, and is then baked at approximately 250°C. This creates a manganese dioxide layer that is the cathode conductor of the capacitor. The contact surface for the manganese dioxide is created by coating the manganese layer with graphite. Finally, the graphite is coated with a silver layer and the cathode connection of the surface mount package is made with silver-loaded epoxy. In the case of through-hole tantalum capacitors, the cathode lead is soldered directly to the silver layer.

Tantalum capacitors typically fail shorted, so fire is a normal consequence of a tantalum capacitor failure. Combustion is aided by the release of oxygen from the manganese dioxide as the capacitor fails. A voltage spike (even within the working voltage range) can precipitate a breach in the dielectric. The capacitor starts to draw current and generates heat, which causes thermal runaway.

Tantalum capacitors have a low failure rate that decreases over time. As long as the capacitors are not stressed, they will have a long life with no wear-out mechanisms as in aluminum electrolytics. ESR in tantalum capacitors decreases with increasing frequency. The ESR is composed of the resistance of the contact material (primarily the graphite) and the manganese dioxide at low frequencies. ESL is not a factor because the inductance is such a small value compared to the ESR. Capacitance and ESR have some temperature-dependence but it is significantly smaller than in aluminum capacitors.

Tantalum capacitors are typically derated by as much as 50% voltage rating to reduce the likelihood of surge failures. Aluminum capacitors can actually withstand current surges and voltage above the rated working voltage. It is not unusual to use a 35 WV tantalum capacitor in a 12 V circuit to minimize surge failures. AVX has a good application note (surgtant.pdf) that describes derating of tantalum capacitors to reduce surge failures.

AVX is manufacturing a line of capacitors based on niobium oxide rather than niobium metal. Niobium oxide is conductive, but niobium pentoxide is an insulator. Niobium oxide has improved reliability and resistance to ignition compared to capacitors made from tantalum or niobium metal. Niobium metal has characteristics equivalent to tantalum metal and is not considered a viable alternative to tantalum. These capacitors may start to see more use because niobium is less expensive and more available than tantalum.

Solid Polymer Electrolytic Capacitors

Solid polymer aluminum capacitors are the superstars of low ESR. These capacitors are similar in construction to aluminum electrolytics. They only have a maximum working voltage on the order of 25 V, so they have a limited range of applications compared to liquid electrolyte aluminum capacitors. These capacitors are also called organic electrolytic capacitors. Solid polymer tantalum capacitors are also available that replace the manganese dioxide with a polymer electrode. A major advantage is that solid polymer electrolytic capacitors are more suited to reflow temperatures than liquid electrolyte capacitors.

The anode of aluminum polymer capacitors is formed from etched aluminum foil with an oxide coating, in the same manner as liquid electrolyte aluminum capacitors. A conductive polymer fills the surface of the aluminum foil and solidifies. Then the polymer is covered with graphite and then a silver layer, in the same way the cathode of the solid tantalum is produced. The ESR is extremely low because the polymer has conductance 10,000 times that of liquid electrolyte and 1000 times that of manganese dioxide. The polymer can withstand high temperatures of 125 C° in modern capacitors. The capacitors fail at high temperature because moisture trapped in the capacitor reacts with the aluminum to form aluminum hydroxide and creates a high resistance in series with the polymer. Manufacturers have developed ways to minimize the moisture and increase the reliability of these capacitors. There are numerous ways that manufacturers configure the foil in an aluminum capacitor. However, none of the manufacturers provide details of construction in their application notes.

Failure rate for these capacitors is increased with both temperature and high humidity. This capacitor type fails open, as does an aluminum electrolytic. An advantage over solid tantalum capacitors is that the materials are not readily flammable.

The slug of a solid polymer tantalum capacitor is formed the same way it would be if the cathode were manganese dioxide. Instead of manganese, the liquid polymer is forced into the porous tantalum structure and is solidified to form the cathode. The cathode connection is made with a coating of carbon and then silver.

Solid polymer tantalum capacitors fail shorted, as do the manganese capacitors. However, the failures are less dramatic because the polymer does not support combustion.

Multilayer Ceramic Capacitors

Multilayer ceramic capacitors have been improved to the point that they can have capacitance in tens of microfarads at voltages up to 16 V. Capacitances around 1 μF and 50 WV are available. These capacitors have very low ESR. They would seem to be ideal, but there are a number of parasitic effects that limit their applications.

Multilayer ceramic capacitors are classic parallel plate capacitors with multiple plates attached to each terminal. The dielectric is a ceramic primarily composed of barium titanate. Figure 6-3 shows how the plates are connected to the terminals. The low ESR is a consequence of the numerous plates connected in parallel plus the low resistance of the plate material. ESL is a significant factor in ceramic capacitors because ESR does not dominate the magnitude of impedance at high frequencies. In other types of capacitors, the ESR and capacitance swamp out ESL effects. The inductance of the capacitor is due to the width and length of the plates. Longer plates produce larger inductance. For that reason, some capacitors are designed with the connections on the long side of the surface mount package instead of on the usual short side. This changes a 1206-size surface mount package into a 0612 package. The ESL is on the order of 500 pH to 1 nH for most surface mount packages. This seems like a very small value, but 500 pH and 10 µF will resonate at 5 MHz.

Figure 6-4 shows the impedance of a multilayer capacitor compared with a typical tantalum capacitor. Both capacitors are 10 µF and have about 1.4 nH ESL. The resonant circuit formed by the capacitance and ESL of the ceramic capacitor has high Q, which shows as the narrow dip in impedance at resonance. Above the resonant frequency, the impedance becomes inductive. The tantalum plot shows that the ESR dominates the impedance above 30 kHz. The ESL has no noticeable effect at high frequencies.

There are five types of ceramic capacitors in common use: C0G (formerly NPO), X7R, X5R, Z5U, and Y5V. Table 6-1 shows class I dielectric temperature characteristics and the associated EIA designations.

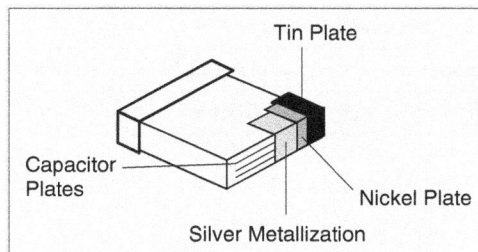

Figure 6-3: Plate connections to terminals in a multilayer ceramic capacitor

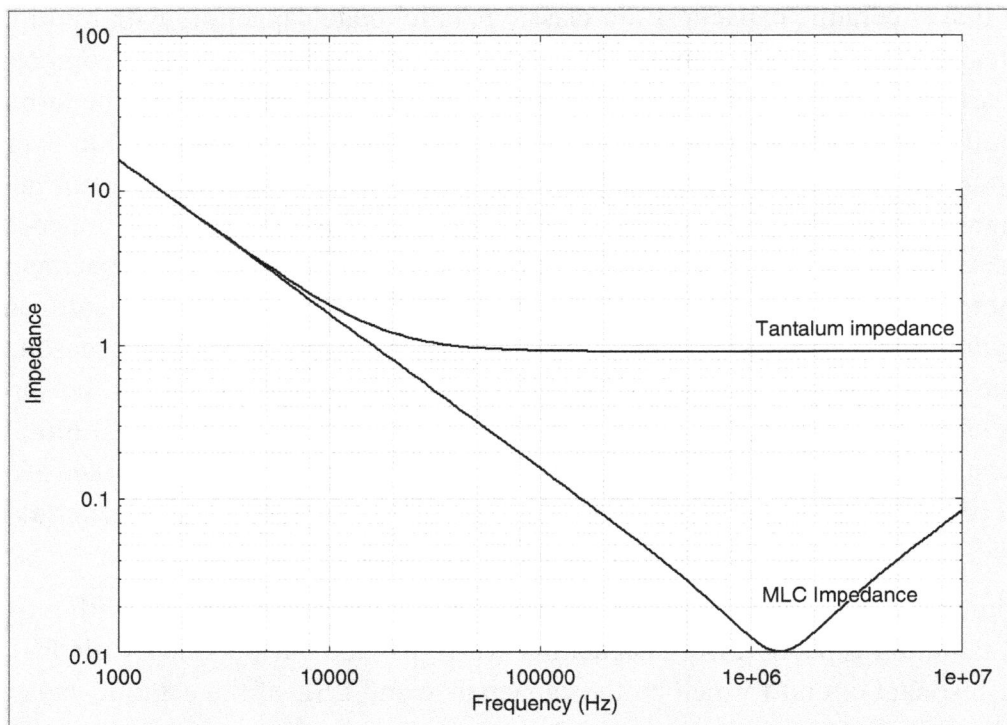

Figure 6-4: Impedance of a multilayer capacitor compared to a typical tantalum capacitor

Table 6-2 shows temperature characteristics of class II and class III dielectrics and the associated EIA designations. Note that both class I and class II, III dielectrics use two letters and a number, but the meanings of the characters are different.

C0G capacitors have the least number of environmental dependencies. The temperature coefficient is zero, the capacitance is not affected by voltage, and

Table 6-1 Class I dielectric temperature codes

Temperature coefficient		Temperature coefficient multiplier		Temperature coefficient tolerance	
PPM per C°	Letter	Multiplier	Number	PPM per C°	Letter
0.0	C	−1	0	± 30	G
0.3	B	−10	1	± 60	H
0.9	A	−100	2	± 120	J
1.0	M	−1000	3	± 250	K
1.5	P	−10,000	4	± 500	L

Table 6-2 Class II, III dielectric temperature codes

Low temperature rating		High temperature rating		Capacitance temperature coefficient	
Temperature	Letter	Temperature	Letter	Percent	Letter
−55°C	X	+ 45 C°	2	± 10.0	P
−30°C	Y	+ 65 C°	4	± 15.0	R
+10°C	Z	+ 85 C°	5	+ 22/−33	U
		+ 125 C°	7	+ 22/−82	V

the dielectric is not piezoelectric. The dielectric constant of Class I ceramic capacitors is relatively low. C0G capacitors typically have values less than 1 nF.

Class II and class III dielectrics have significantly larger dielectric constants than C0G, so the capacitance is significantly larger. Y5V capacitors can have values in tens of microfarads at low voltage ratings. The Class II and Class III dielectrics have multiple shortcomings. The temperature coefficient is quite large. Figure 6-5 shows capacitance change with temperature that is typical of Z5U and X7R capacitors. The capacitance decreases at both ends of the temperature range. Ceramic capacitors can have a significant decrease in capacitance with applied DC voltage. Z5U capacitors can decrease in value by as much as 80%. X7R capacitors have less decrease with DC voltage. The capacitance increases with applied AC voltage. It is important to use the manufacturer's data sheet to verify how temperature and applied voltage will affect your designs. Data sheets do not always specify all parameters that will affect value, so you may need to characterize capacitors to ensure a proper design. Ceramic capacitors also age as they cool over significant periods of time. These capacitors slowly decrease in value as they cool. Heating them, as in reflow or IR soldering, will raise the capacitance again. X7R capacitors are better behaved than Z5U capacitors with respect to aging.

Class II and class III dielectrics are piezoelectric, so mechanical shock can produce voltage. This is referred to as being microphonic. In most applications in power supplies, this effect will not be an issue. The other side of the piezoelectric effect is that applying an AC voltage will cause the capacitor to vibrate. In a power supply, this will not be a problem except where audio frequencies are used for the switching frequency. Class III ceramic has a much larger piezo effect than class II ceramic.

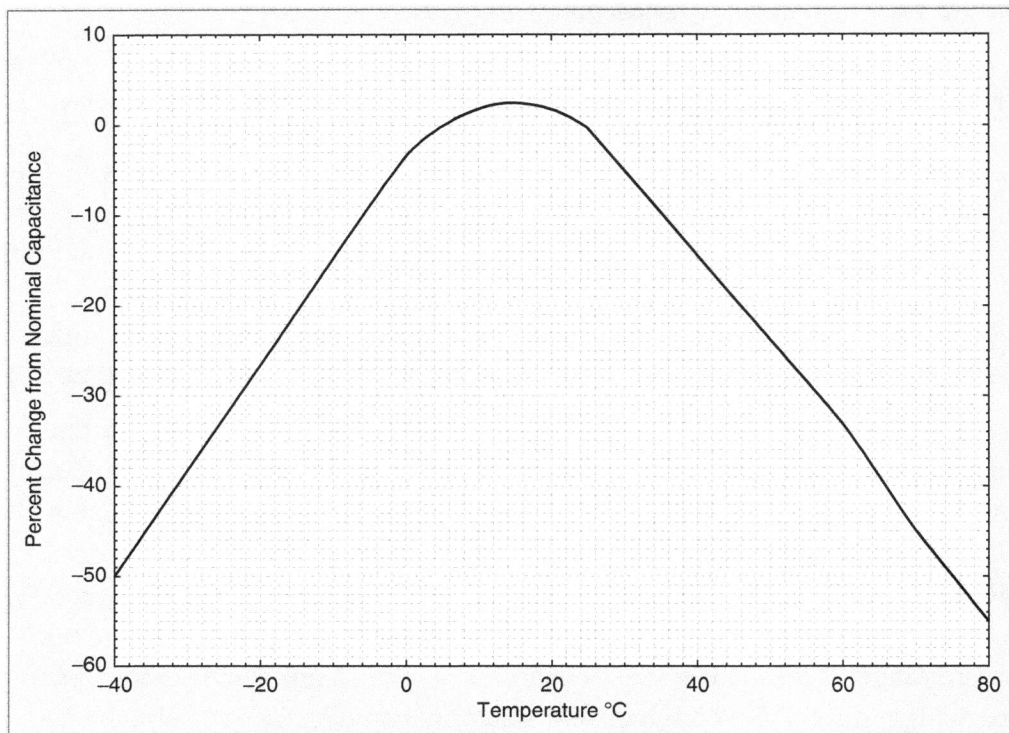

Figure 6-5: Capacitance change with temperature that is typical of Z5U and X7R capacitors

Ceramic capacitors have excellent ripple current capability for a given volume. They dissipate proportionately less power than polar capacitors because ESR is so low.

Film Capacitors

Film capacitors excel in high current AC applications such as coupling capacitors in bridge designs. They can also replace aluminum electrolytics in some applications. CDE manufactures a line of film capacitors that can replace aluminum electrolytic capacitors in input filter supplies with higher ripple ratings, greater life ratings at high temperature, and lower ESR. They accomplish all those parameters in a volume roughly equal to that of aluminum capacitors.

Film capacitors come in a variety of configurations. The film can be polyester (Mylar®), polypropylene, polycarbonate, polyethylene napthalate (PEN), or polyphenylene sulfide (PPS). The plates can be either foil or metallization deposited on the film. Foil construction is limited to smaller capacitance values because the thickness of the foil is equal to the thickness of the film. Metallized capacitors allow much larger capacitances in the same volume because the metallization adds very little thickness to the film. Metallized capacitors can self-heal as described in Chapter 3, but foil capacitors do not. Through-hole parts are manufactured by rolling the film and plates into a cylinder. The leads are connected to opposite sides of the rolled assembly. High current devices typically have the leads welded to the plates. Surface mount devices are created as stacks of plates and dielectrics in much the same manner as multilayer ceramic capacitors. The biggest drawback to surface mount film capacitors is that they have limited capacitance values due to their small size. Film capacitors have very low ESR, but they are much larger than polarized capacitors for the same capacitance.

Polyester film is rated from −55 C° to 85 C°, or to 125 C° with derating. Polypropylene is rated from −55 C° to 85 C°, or to 105 C° with derating. Polyester and polypropylene are standard dielectrics for through-hole parts. PEN and PPS films are used for surface mount devices in order to withstand the temperatures of reflow soldering. Polyester and polypropylene parts would melt at reflow temperatures.

Film capacitors have very low temperature coefficients. They also have very tight tolerances on capacitance value compared to the polarized capacitors. Film capacitors are available in voltage ranges from 50 V to thousands of volts. The minimum thickness of the film limits the lowest working voltage to 50 V. It would be difficult to manufacture capacitors with thinner films.

Film capacitors find use in timing circuits where the capacitance value stability is important. They are also used extensively in switchmode supplies, in snubber and clamp circuits. Again, the stability of the capacitance value is a prime consideration.

Resistor Characteristics

Resistors come in a wide variety of configurations. As with capacitors, certain configurations work better than others, depending on the application. We'll look at the variations of surface mount and through-hole parts.

It is important to accurately determine the maximum current or voltage for a resistor so that you can choose the wattage rating that is appropriate. Since power is proportional to either current squared or voltage squared, even small errors that understate the actual value can cause a problem with the size. An error of + 50% in voltage or current will cause the actual power to be larger by 125%. Metal resistors have a positive temperature coefficient. Carbon is a semiconductor so it has a negative temperature coefficient, as do silicon and germanium. The power rating for a resistor is typically the power that can be dissipated when the resistor is below 70°C. The applied power must be reduced at higher temperatures.

A characteristic of resistors that is not covered well in engineering schools is the voltage rating. Each resistor case has its own maximum voltage rating. Film and carbon composition resistors typically have working voltage in the range of 200–350 V. This is very important in applications on the input side of off-line power supplies. Surface mount resistors typically have working voltage ratings of 50–150 V. In higher voltage applications, you would want to break the resistance into two resistors in series to obtain the necessary voltage rating.

We only looked at current mode controllers in Chapters 4 and 5. A very low resistance current sensing resistor is required for all current mode control circuits. The current sense resistor in high current "point of load" applications, such as Pentium-class CPUs, may need an extremely small value because the currents can approach 50 A. The first requirement is for the resistor to have the smallest inductance possible. A resistor with 10 nH inductance will generate 1.0 V with 50 A change in 500 ns.

Current levels in tens of amps will generate voltage across even very wide circuit traces. This makes it important to pay attention to the layout of the voltage sensing traces. Current sense resistors can be obtained in a four-terminal Kelvin voltage sensing configuration so that the voltage measurement is as accurate as possible. A Kelvin connection is a second set of terminals on the resistor that are intended for connection to a voltage measurement; they are not intended to carry current. This allows the voltage sense to measure the current directly across the calibrated resistance. Two terminal resistors are sensitive to the layout of the circuit. The solder to hold the resistor to the circuit board and the size of the pad can materially change the voltage sensed when dealing with resistances below 0.005 Ω.

All resistors have ESL. A resistor that is classified as noninductive is manufactured in a manner that limits inductance to the smallest practical value.

Carbon Composition Resistors

Carbon composition resistors are probably the oldest variety, dating to the earliest days of electronics. These resistors are only available as through-hole parts. They are manufactured in 1/8 through 2 watt sizes with tolerances generally ± 5%, ± 10%, or ± 20%. The core is molded from carbon and a binder to create the desired resistance. The core has a wire contact with a cup-shaped connection to hold the resistance slug at each end. The entire core is molded inside of an insulating body, which is usually similar to bakelite. The body is porous, so carbon composition resistors are sensitive to the relative humidity of the environment. They will absorb moisture, which can change the resistance value over time. The primary advantage of carbon composition is that it is noninductive. These resistors are also somewhat tolerant of short-duration pulses that exceed the power rating. Long-term exposure to heat and excessive dissipation will cause a chemical change in the slug that permanently increases the resistance.

Carbon composition resistors cause significant electrical noise that becomes worse as temperature rises. Noise is also more of a problem in the higher resistance values. This is noise similar to shot noise in semiconductors. Carbon composition resistors are available from only a few manufacturers because of the problems associated with moisture absorption, noise, and large tolerances. They are primarily useful in switchmode supplies for RC snubbers and for EMI filter applications on the input supply.

Film Resistors

Film resistors are manufactured by depositing either a thin film or thick film of resistance material on a substrate. It is common to have each resistor trimmed by the manufacturing machinery using a laser to adjust the resistance to the desired value. Through-hole resistors are trimmed by creating a spiral cut in the film on the ceramic or glass substrate. This creates an inductor that can have significant ESL. There are film resistors that are specified to be noninductive that are trimmed in a manner that minimizes the inductance. Another consequence of spiral trimming is that it increases the parasitic capacitance of the resistor. Surface mount resistors are manufactured with a continuous strip of resistance material between the terminations. One method of trimming these resistors is to have the laser cut partly across the strip to reduce the effective width and to increase the effective length. This creates the equivalent of a printed inductor. The inductance is much less than that created by the spiral method of through-hole parts.

It is important to read the manufacturer's data sheet for the technology you select to ensure that the noise performance is adequate for your application. Different film technologies have varying noise versus temperature characteristics. This is most likely to be an issue for the voltage divider circuit of the control IC. It is also important to determine the pulse and overload characteristics from the data sheet if you choose a resistor that will operate near its rated power limit. Ability to absorb power transients varies among technologies. Film resistors are available in varying value tolerances, with 1% or better available at very reasonable prices. A major advantage of 1%- or 0.1%-tolerance film resistors is the fine granularity of resistance values. It is reasonable to expect that you can design a circuit like the output voltage divider using standard resistance values rather than using a potentiometer and labor to set the output voltage.

Film resistors are available in numerous high power configurations. The mounting methods mirror those for semiconductors. Examples include TO-220 power tab, DPak, and flange mount similar to RF power transistors. Resistors with these types of mounting can dissipate tens of watts and conduct tens of amps.

Film resistors are available in very low resistances for current sense applications. These typically have 1- or 2-watt ratings and are available in both surface mount and through-hole configurations.

Wire Resistors

Wire resistors come in a variety of configurations. Very high wattage resistors are usually wire-wound, which creates a large amount of inductance. Such

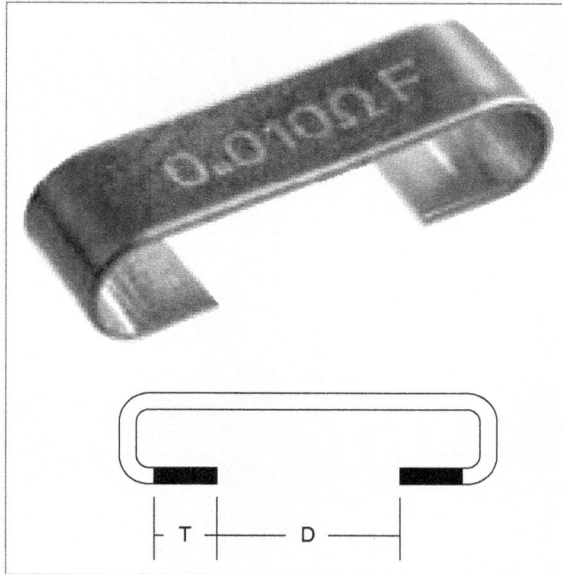

Figure 6-6: Configuration used by some manufacturers for current sense resistors in the 0.010–0.0005 Ω range

resistors are wound using high resistance wire on a ceramic tube and encapsulated in a ceramic or porcelain insulation. These resistors are of use only on the primary side of off-line supplies because of their large ESL.

Current sense resistors can also be manufactured as small metal strips that can be surface mounted. Figure 6-6 shows a configuration that is used by some manufacturers for current sense resistors in the 0.010–0.0005 Ω range. This configuration has very low ESL and excellent high temperature characteristics. The area designated by T in the side view represents the area that must be in contact with the pads on the PCB in order for the resistance value to be proper.

CHAPTER 7

Semiconductor Selection

- Diode Characteristics
- Junction Diodes
- Schottky Diodes
- Passivation
- Bipolar Transistors
- Power MOSFETs
- Gate Drive
- Safe Operating Area and Avalanche Rating
- Synchronous Rectification
- Sense FETs
- Package Options
- IGBT Devices

Semiconductor Selection

Diodes are manufactured in a variety of technologies and materials. We will look at the strengths and weaknesses of a wide variety of diode technologies based on application in switchmode power supplies.

Diode Characteristics

In engineering school, we learned that a diode will conduct when forward biased (with a more or less constant voltage drop of 0.7 V) and it will stop conducting as soon as the voltage falls below this forward voltage. This is a reasonable approximation for the slow circuits we dealt with in school, but it falls flat when looking at fast rise and fall times and high frequencies in switchmode power supplies.

Junction Diodes

"PN" junction diodes consist of a piece of P material (current flow is from holes) on one terminal and a piece of N material (current flow is from electrons) on the other terminal. The area between the two doped regions is an area called the depletion layer, where very few holes or free electrons exist. The depletion layer acts very much like intrinsic semiconductor material. When the PN junction is forward biased, holes and electrons are injected into the depletion layer where they recombine and allow current to flow. The thickness of the depletion layer decreases to a very small size when forward biased. Likewise, when the PN junction is reverse biased, the applied voltage pulls holes from one side and electrons from the other side of the junction. This increases the thickness of the depletion layer. The depletion layer acts as a dielectric and the P and N regions act as the plates of a capacitor. A PN junction diode has capacitance when forward biased and when negative biased. The capacitance decreases as the reverse bias is increased.

There is a finite time required to inject enough charge into the depletion layer to cause current to flow through the diode when turning on the diode. This is called forward voltage recovery time. The voltage increases beyond the normal conduction voltage for a very short period of time. The forward recovery voltage is usually a very small value and the time is usually not critical in switch-mode power supplies. Forward recovery time is usually ignored when selecting a diode. Low forward recovery voltage is usually a byproduct of a low reverse recovery time. Most data sheets do not specify forward recovery time.

Holes are minority charge carriers and move very slowly compared to electrons. As holes cross the depletion layer, they combine with electrons and are destroyed. When the applied voltage across the diode decreases to zero and goes negative, injection of holes and electrons ceases. However, there is a finite amount of time needed for all of the holes in the depletion layer to recombine with electrons. The amount of time for current to stop is called the reverse recovery time. The recombination of holes and electrons causes reverse current to flow for a very small amount of time after the applied voltage goes to zero and becomes negative.

Reverse recovery time is the first selection criterion for choosing a diode. The combination of forward recovery and reverse recovery must be much smaller than the shortest duty cycle in order for the diode to operate primarily in conduction mode. Junction diodes are classified as standard recovery, fast recovery, or ultra-fast recovery.

Standard recovery diodes are only suitable for power line frequencies because the reverse recovery time is typically 1 μs to perhaps 10 μs. Most standard recovery diode data sheets do not even specify reverse recovery time. Fast recovery diodes typically have reverse recovery times in the 100–250 ns range, and ultra-fast diodes typically have reverse recovery times in the 25–50 ns range.

Power diodes are actually produced using a PIN (P-Intrinsic-N) structure rather than a classic diffused PN junction. The intrinsic silicon between the P and N layers improves the reverse voltage capability of the diode. PIN diodes are available in voltage ratings to more than 1000 V. The "intrinsic" layer is actually not intrinsic but, rather, is very lightly doped N material. A thicker intrinsic layer will increase the blocking voltage, but it will also increase the forward voltage because of the resistance of the intrinsic material. The intrinsic layer can be doped with gold to increase the rate of minority carrier recombination so reverse recovery time is smaller. However, there is a trade-off between lowering reverse recovery and increasing forward voltage. Forward voltage for a PIN diode is typically on the order of 0.8 V. PIN diodes have reasonable reverse leakage current that is only moderately affected by increased temperature. Most ultra-fast PIN junction diodes are manufactured as fast recovery epitaxial diodes (FREDs). This process is similar to the epitaxial process used to manufacture bipolar transistors. Figure 7-1 shows a cross section of a FRED. This process allows a manufacturer to greatly reduce the reverse recovery time and the reverse recovery current. Philips and International Rectifier have good application notes on their websites that discuss FREDs in detail.

Reverse recovery current is responsible for several undesirable effects in switchmode power supplies. Large reverse currents will cause problems for the switch because the switch will have to conduct the reverse current as well as

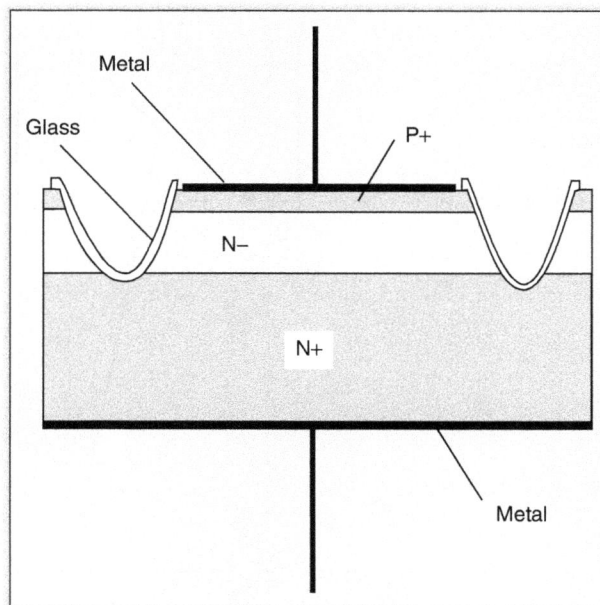

Figure 7-1: Cross section of a fast recovery epitaxial diode (FRED)

the current required to charge the inductor. The waveform of the reverse recovery current contributes to RFI from the power supply. Figure 7-2 shows the waveforms for reverse recovery.

The current through the diode is constant in our example, while the diode is forward biased. When the switch opens, the current through the diode begins to decrease, but the forward voltage continues because the diode is still conducting. The current eventually becomes negative as the excess minority carriers begin to recombine with electrons. The voltage is still positive but lower than the forward voltage. This time is referred to as t_A by EIA convention. As soon as the depletion region begins to block the applied voltage, the recovery current begins to decrease during time t_B. Very short t_B times are referred to as snap recovery. This effect creates large amounts of RFI. FREDs are designed to reduce the reverse recovery current (by limiting the total amount of minority

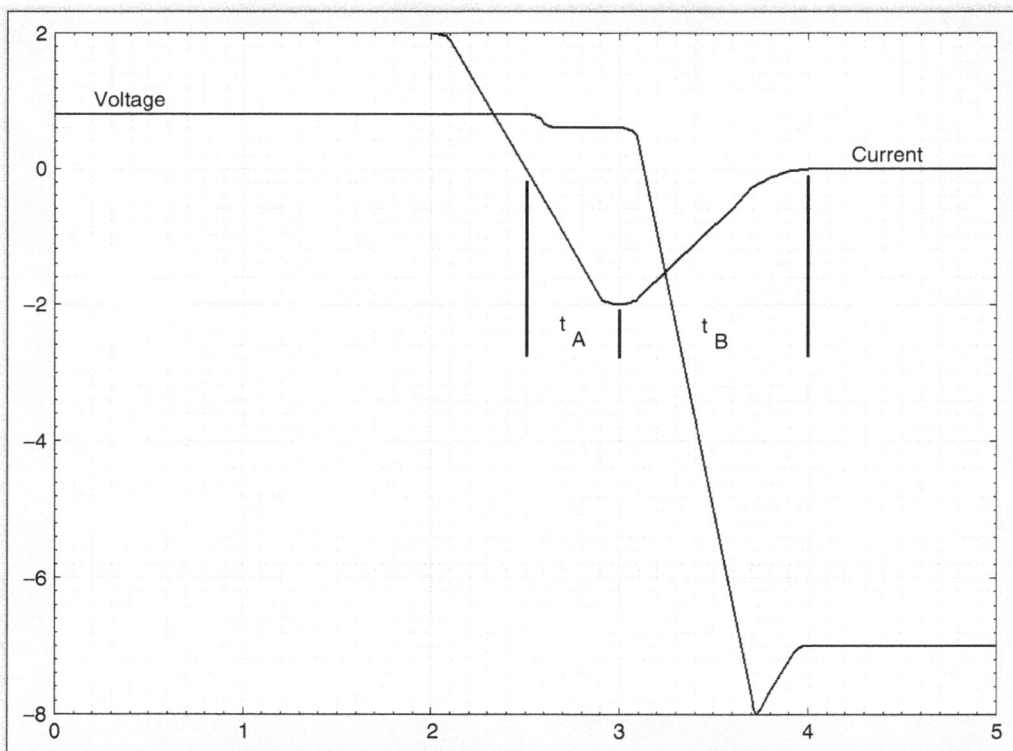

Figure 7-2: Waveforms for reverse recovery

carriers) and to have a soft recovery characteristic. EIA defines the softness factor as t_B/t_A. A value greater than 1 is considered soft recovery and a value less than 1 is considered snappy. Figure 7-3 shows a problem with using this softness factor. Each of the waveforms has the same softness factor, but the maximum slope of the left waveform is smaller and will have a smaller number of harmonics than the other two. Most FRED manufacturers have worked to reduce reverse recovery current and to provide a very soft recovery waveform. If your application is sensitive to RFI, you will want to choose a modern FRED specified for softness. Softness and total reverse current are areas of heavy competition between manufacturers. FREDs have several advantages over standard ultra-fast recovery diodes (usually manufactured using a double diffused process):

1. The reverse recovery is significantly lower than standard diodes because the stored charge in the depletion/intrinsic layer is smaller.

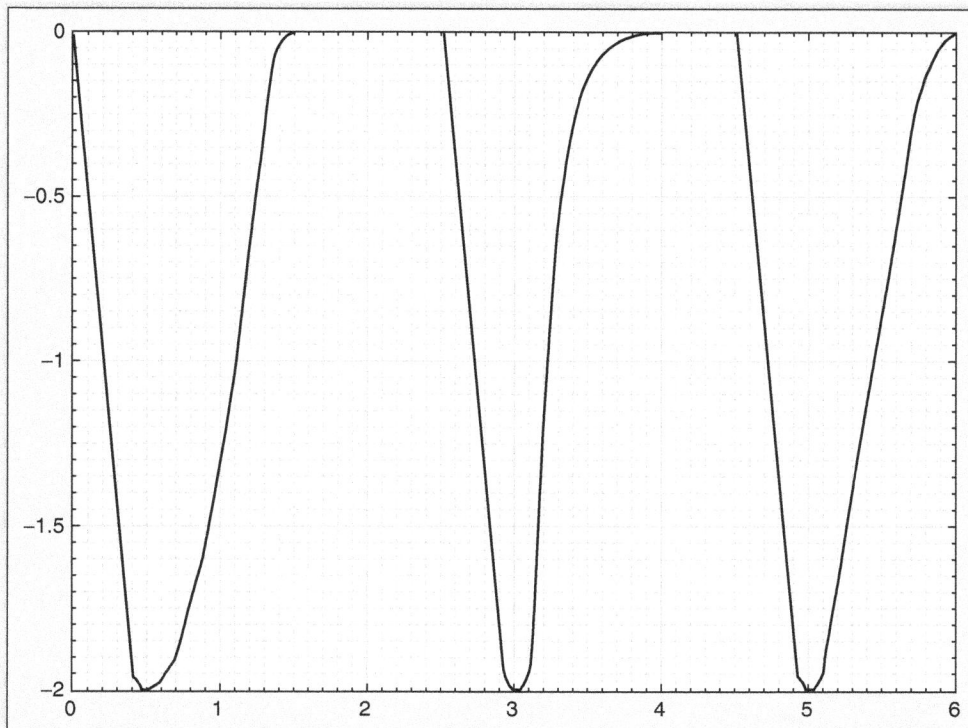

Figure 7-3: Three waveforms with identical softness factors but differing slopes

2. FREDs have significantly softer recovery than standard diodes. The softness factor is at least twice to ten times that of standard diodes.

3. Reverse recovery time is less temperature-dependent than standard diodes.

4. V_F is typically lower because the epitaxial process gives better control of the doping of the N and P layers.

The FRED is designed to have a minimum stored charge that is typically on the order of 100 nC. However, even such a small amount of charge will create current of many amps when it flows in only 20 ns. Reverse current of 2–8 A is a reasonable expectation.

We looked at RC snubber circuits in Chapter 5 to reduce the effects of reverse recovery in rectifiers. The snap effect of hard recovery diodes is responsible for exciting the combination of the diode junction capacitance and stray inductances in the circuit. Soft recovery diodes can eliminate the need for snubbers across the rectifiers, since the softer waveform will not deliver as much energy to excite the tuned circuits.

Reverse recovery in the power line rectifier diodes can contribute to RFI. The reverse recovery in a junction or standard recovery PIN diode can be on the order of tens of microseconds and contain thousands of nanocoulombs of charge. Switching from standard recovery diodes to ultra-fast soft recovery diodes can reduce the energy that must be filtered by the input EMI filter. Reverse recovery of standard recovery diodes can generate harmonics of the power line frequency up to 10 MHz or more.

Schottky Diodes

The Schottky effect occurs when metal is in contact with semiconductor material. The very oldest diodes ("cat's whisker" diodes) used metal with a sharp point in contact with a semiconductor material. The metal in contact with the semiconductor material creates a space charge region that allows current to flow in one direction but not the other. Schottky diodes are an extension of that technology. Modern Schottky diodes have a structure as shown in Figure 7-4.

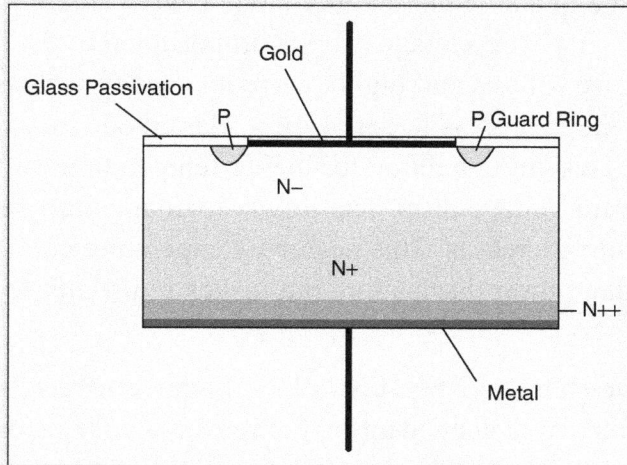

Figure 7-4: Structure of a modern Schottky diode

The rectifying junction is created with a layer of metal (usually gold, platinum, aluminum, or palladium) deposited on the surface of a lightly doped semiconductor. The metal and the doping level affect the rectifying characteristics. The rectification property occurs because of the difference in energy levels of the materials. The back side of the semiconductor is more highly doped with a metal contact. The rear contact is called an ohmic contact because the energy levels of the materials are very close, so it looks like a resistor. Current flows in a Schottky diode because electrons in the metal exceed the barrier potential when the junction is forward biased. This is why Schottky diodes are also called hot carrier diodes.

Current flow in the semiconductor material is composed of electrons. Electrons are majority carriers, so current flow is faster than the current in the P material in junction diodes. Majority carrier current flow makes Schottky diodes the fastest of all diodes. Since there is no excess of minority carriers in the junction, the diode turns off as soon as the applied voltage goes to zero. There is a very small amount of reverse current that flows due to charging the junction capacitance. This capacitance is extremely small, so the reverse current is also extremely small. Schottky diodes have essentially no forward or reverse recovery times because conduction does not depend on minority carriers.

The forward voltage of a silicon Schottky diode is very small. It is typically on the order of 0.2–0.45 V. The voltage drop is proportional to the blocking voltage. Higher blocking voltage and higher current capacity increase forward voltage drop because the N layer is larger. A 10 V class diode may only drop 0.3 V when conducting. Designing a diode for higher temperature capability increases the forward voltage drop. The forward voltage drop decreases as the junction temperature decreases. This negative temperature coefficient is advantageous in managing power dissipation, but makes it difficult to use the diodes in parallel.

The major negative characteristic of Schottky diodes is reverse leakage current. Reverse leakage current is dependent on both reverse voltage and junction temperature. Leakage increases with increase of both temperature and reverse voltage. The reverse leakage increases exponentially with temperature. The maximum leakage is also very dependent on the manufacturing process. Devices rated for higher reverse voltage and higher maximum junction temperature have lower leakage than lower voltage and lower temperature parts.

Manufacturers have been steadily increasing the reverse voltage capability of Schottky diodes. Ten years ago, Schottky diodes were capable of handling reverse voltage for 5 V or 12 V outputs only. They are now available in gallium arsenide and silicon carbide technologies with higher voltage ratings. Maximum PRV for silicon Schottky diodes is on the order of 150 V, which makes them suitable for 48 V telecom universal input power supplies. GaAs diodes have PRV ratings as high as 300 V, which allows use in outputs close to 100 V. The forward voltage is typically 0.8 V for GaAs Schottky diodes. This is not usually a problem because the current is typically much smaller in high voltage supplies.

Until recently, the only choices for the diode in boost mode power factor controllers were high voltage ultra-fast diodes or FREDs. The forward and reverse recovery of these diodes limited their use to frequencies from 100 kHz to perhaps 300 kHz. SiC Schottky diodes are available from Advanced Power Technologies, Infineon, and Cree with PRV ratings up to 1200 V. The manufacturers expect future generations of SiC diodes to have PRV ratings up to 2000 V. The typical forward voltage for SiC diodes can be as high as 1.5 V for 600 V diodes and 3.0 V for 1200 V diodes, so power dissipation can be higher than that of FREDs for the same current level. A significant portion of loss in FREDs is due to reverse recovery, whereas SiC Schottky diodes have primarily conduction losses. This is not likely to be an issue since SiC diodes can withstand significantly higher temperatures than silicon diodes for the same physical size of die. Forward voltage has a positive temperature coefficient, so it is possible to parallel SiC diodes for higher current capability. SiC diodes also have the advantage that the reverse leakage current has a much smaller temperature dependence than silicon or GaAs diodes.

Since Schottky diodes have no forward or reverse recovery, SiC diodes allow power factor correction circuits to operate above 500 kHz. These higher frequencies allow for much smaller inductors and higher efficiency. EMI filtering is improved because the components can be much smaller for the same amount of attenuation.

Passivation

Any semiconductor that is intended for high voltage use will require passivation. The voltage gradients at the edges of a chip can be very large at the sharp edges of the chip. The large voltage gradients lead to unwanted reverse leakage current. Passivation is applied around the high voltage active area of a semiconductor device to electrically isolate the edges. Figure 7-5 shows passivation on a FRED, a Schottky diode, and a bipolar transistor.

Planar passivation involves using both a guard ring as well as glass deposited on the surface of the chip. Mesa passivation uses a trench to improve the voltage withstand properties on the surface of the chip. Both processes are used, depending on the manufacturer.

Bipolar Transistors

Figure 7-5 shows the cross section of a high voltage bipolar transistor. The large N– region between the base and the collector is similar to the N– region in a PIN junction diode. Its purpose is to increase the voltage capability of the transistor. Just as in the diode, increasing the thickness of the N– layer

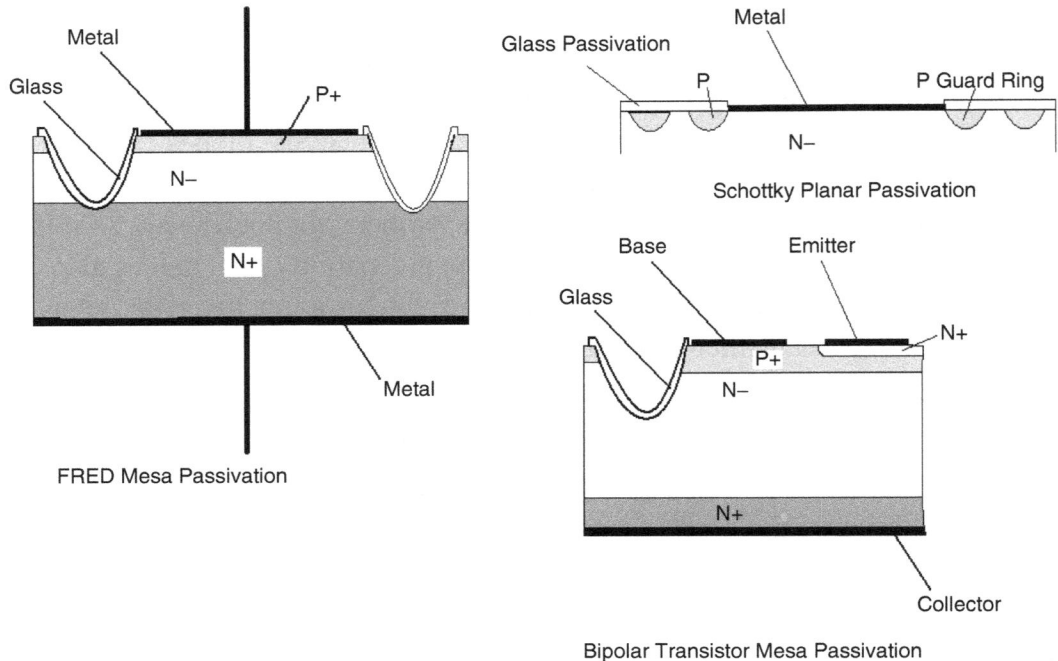

Figure 7-5: Passivation on a FRED, a Schottky diode, and a bipolar transistor

increases voltage capability. The increased voltage comes at the expense of decreasing gain at high currents and increasing the switching time.

A transistor rated at 1500 V V_{CES}, as would be used in a forward converter, will be much slower than a 400 V transistor. Turn-off time is the most significant parameter for a bipolar transistor. Turn-on is relatively straightforward and is a direct result of the injection of base current causing collector current to flow. Eventually the collector-emitter voltage saturates and the collector voltage is below the base voltage. This forward biases the collector-base junction which adds more minority carriers. Turn-off takes considerably longer because the minority carriers must be swept out of both diode junctions in order to turn off the transistor. This is similar to the reverse recovery described for junction diodes. It is normal practice to use a small amount of negative bias at the emitter-base junction to force negative current to enhance elimination of minority carriers.

Collector voltage ratings for transistors can be quite confusing. The BUT11 is a common switchmode transistor that is rated for 450 V V_{CEO} but is rated for

1000 V V_{CES}. The V_{CES} rating applies to the transistor when the base is shorted to the emitter or when it has a negative voltage applied. V_{CEO} applies when the base is open circuited. A small amount of leakage current always flows in a transistor with a high voltage applied. Some of this current is holes that are generated in the base region. These holes are swept out of the base region if the base is shorted to the emitter or if the voltage is negative. These holes will drift toward the emitter if the base is left open circuit. The base leakage current acts like base current and turns on the transistor. The V_{CEO} rating is typically half of the V_{CES} rating.

Safe operating area and secondary breakdown are major design concerns for bipolar transistor circuits. At high collector voltage and high collector current, the voltage gradient in the collector region can become great enough to lead to avalanche multiplication of current. The avalanche current crowds in the base region and creates very high temperatures, which quickly destroys the transistor. This effect is called secondary breakdown. The transistor data sheet always gives a forward biased safe operating area (FBSOA) graph for various pulse times. Forward biased refers to the base-emitter junction being forward biased. The right edge of the FBSOA graph is the area that can produce secondary breakdown. Power transistor data sheets also frequently give a reverse biased safe operating area (RBSOA) graph that describes the safe operating area when the emitter-base junction is reverse biased but collector current is still flowing. The RBSOA is significantly smaller if the base voltage is zero. If a negative base voltage is applied, the RBSOA tails off and allows more area below the curve, as shown in Figure 7-6. The smaller RBSOA for zero base current creates design problems for switchmode circuits during turn-off, so switching circuits usually employ reverse base bias for turn-off. Snubber circuits are frequently employed to guarantee both FBSOA and RBSOA operation. It is important to note for our example transistor that the FBSOA limits collector current to values significantly below the maximum saturated current when voltage above 20 V is applied. Collector current derating is necessary and is a significant drawback to bipolar transistors.

Optimum drive of a high voltage bipolar transistor requires more than a simple square current waveform. In order to minimize switching losses, the base must

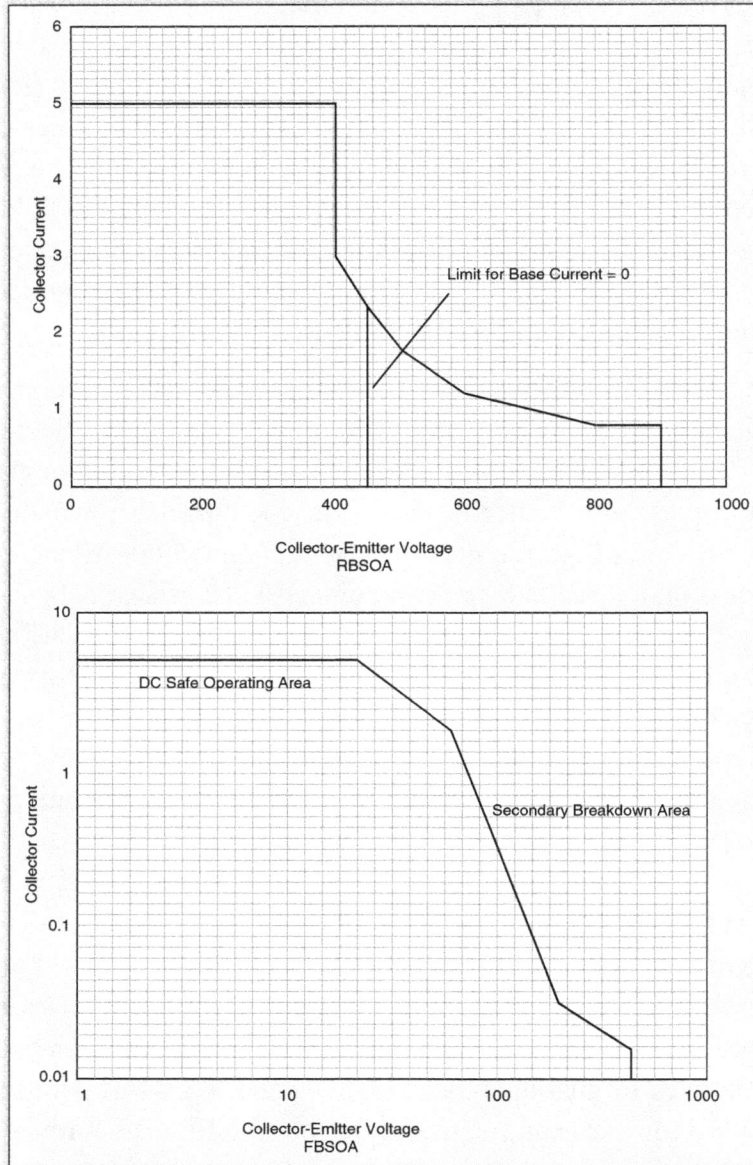

Figure 7-6: Graphs for a reverse biased safe operating area (RBSOA) and a forward biased safe operating area (FBSOA)

be overdriven. The process that lowers the collector resistance injects charge into the lightly doped N− region. The faster this charge is injected, the faster the transistor turns on and transitions to saturation. Figure 7-7 shows base cur-

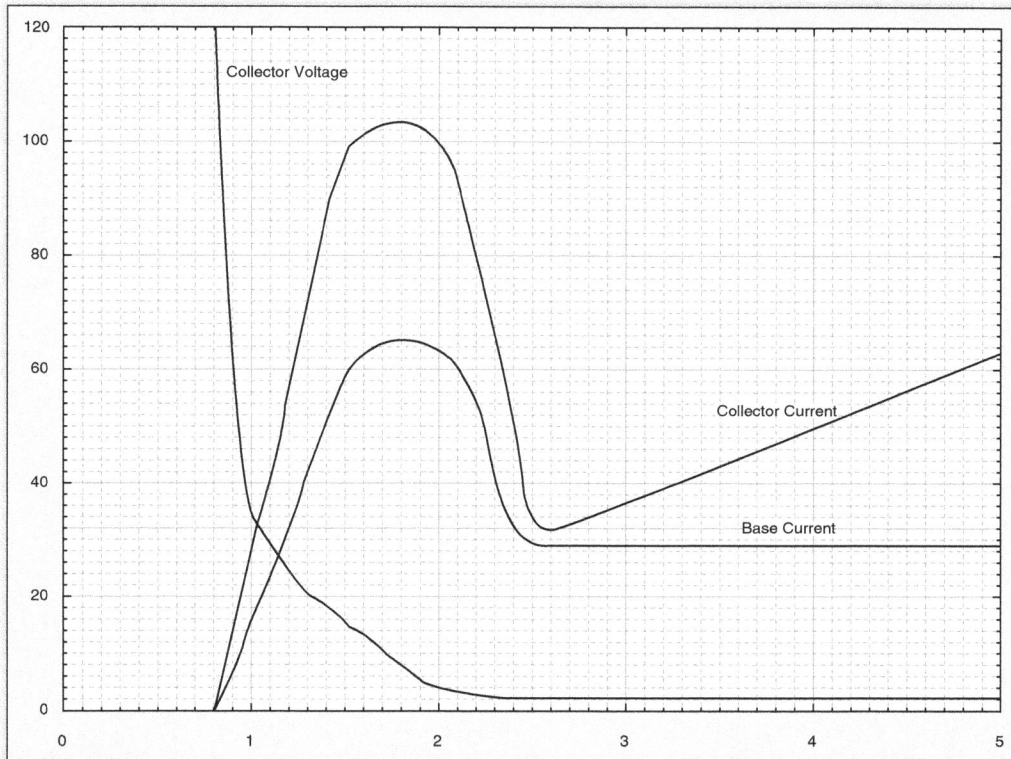

Figure 7-7: Base current, collector current, and collector voltage during fast turn-on

rent, collector current, and collector voltage during fast turn-on. Without the base current pulse at the beginning of drive, the collector voltage will remain substantial while the collector current builds up and creates significant turn-on loss. As a first approximation, the turn-on pulse should equal the storage time. This is reasonable, since we are trying to inject enough charge into the collector to saturate the transistor.

Likewise, simply removing base drive during turn-off is not sufficient to minimize switching loss. Turning a transistor off too quickly or too slowly can each lead to turn-off losses. Turning off the transistor too quickly will shut off the collector base junction abruptly and trap charge in the collector. The trapped charge will recombine slowly, which increases the time for collector current to go to zero. If the transistor turns off too slowly, the collector voltage will rise to a substantial value while collector current is still flowing.

The storage time is increased slightly if we drive the transistor off with an inductor in series with a negative voltage. The combination of voltage and inductance matches the reverse current to the storage time of the device. Figure 7-8 shows optimum turn-off waveforms. The optimum negative base current is about one-half of the collector current. Notice that the collector current does not start falling until the reverse base current has peaked. This occurs because the collector current is split between base current and emitter current. Collector current begins to fall when emitter current goes to zero.

These waveforms are very similar to the soft reverse recovery waveforms we saw with FREDs. Turn-off dissipation is also minimized by using a rate-of-rise snubber on the collector to delay the rise of collector voltage. The Miller charging current that occurs with high collector dv/dt opposes the current trying to turn off the transistor. The snubber also minimizes this parasitic effect.

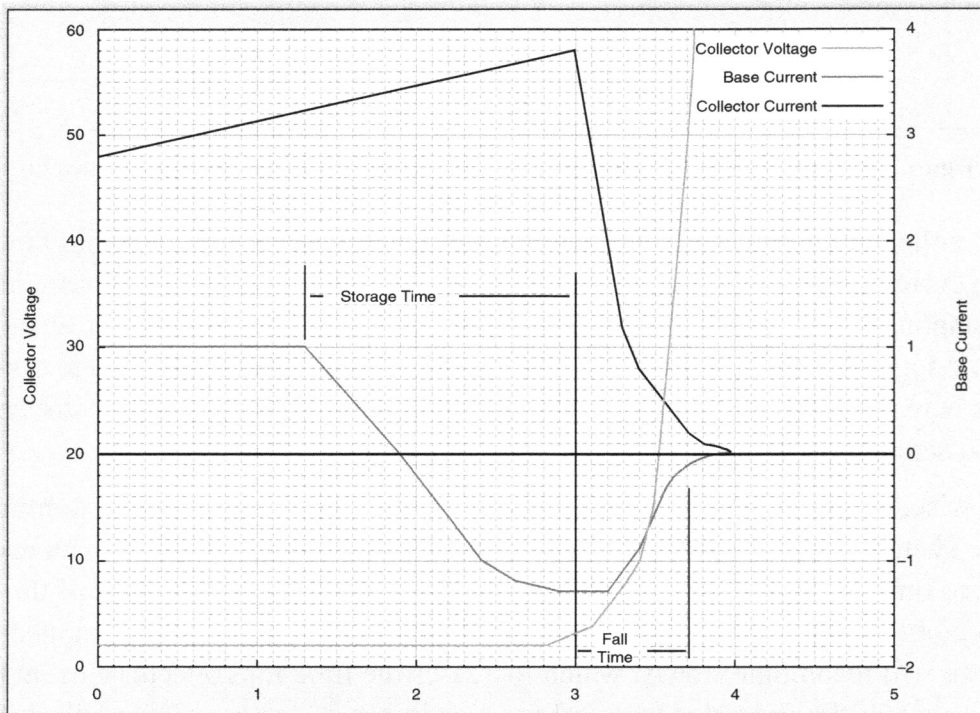

Figure 7-8: Optimum turn-off waveforms

Figure 7-9 shows an optimized transistor drive circuit. T1, Q2, and Q3 are necessary for level shifting the drive signal from Q4 and Q5, which would typically be inside the control IC. T1 shifts the 0 V to 5 V drive signal to ± 5 V to the driver. D3, C2, and R2 provide the turn-on current. The rule of thumb is that the R2/C2 time constant should match the storage time of the transistor. R2 is chosen to give the desired base current with 5 V of drive voltage. D3 ensures that the charge on C2 does not affect turn-off. L1 provides storage time control for turn-off. It is typically on the order of 1–10 μH. Diode D1 ensures that the inductor only acts during turn-off. The rate-of-rise snubber (C1, D2, R1) assists in keeping the voltage inside the FBSOA (20 V for the BUT11) while the collector current is high.

Turn-off time limits most bipolar transistors to use below 100 kHz, but 50 kHz is a more practical upper limit. Bipolar transistors are still competitive with MOSFETs below 50 kHz and at 400V or lower. Saturation voltage for the BUT11 (a very common high voltage transistor) is below 1 V in normal operation, so conduction losses are quite low. They cost about $0.50 each in quantity. Even with the extra drive components, the total bill of materials can be less than that of an equivalent MOSFET. As voltage or frequency requirements go up, bipolar transistors are no longer competitive. There is only a small number of bipolar device types available above 400 V, and bipolar transistors cannot

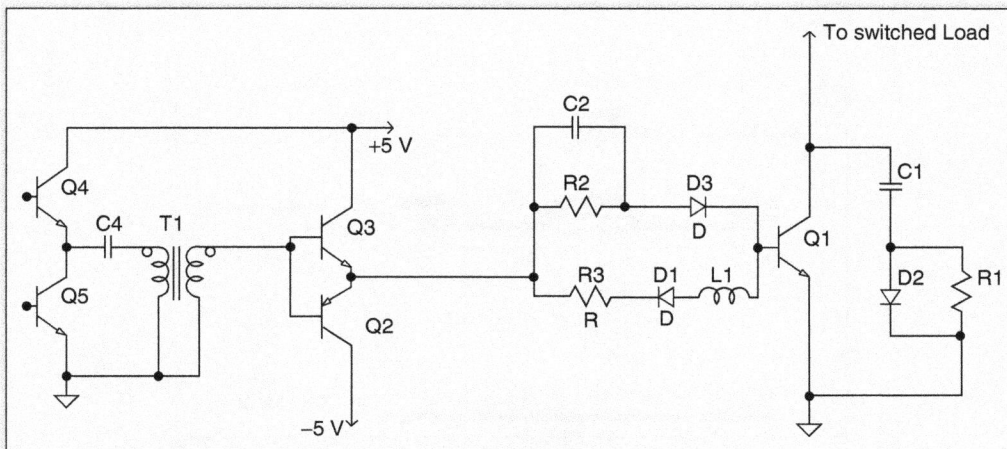

Figure 7-9: Optimized transistor drive circuit

switch high voltage at high frequencies. The trend to ever-higher switching frequencies has essentially eliminated bipolar transistors from mainstream use.

Power MOSFETs

Figure 7-10 shows a cross section of a vertical MOSFET. Once again, the structure is similar to other high voltage devices. Current flows vertically through the chip, and the voltage rating is a factor of the large N− region.

The chip is fabricated in a process similar to that of a FRED. Once the chip epitaxial layers are created, the gate oxide is formed on the surface of the chip. The polysilicon gate material is deposited and then the final silicon oxide is formed over the gate material. The last step is to deposit the aluminum source connection over the source area of the chip. Figure 7-10 only shows two cells of the MOSFET. The whole MOSFET is composed of thousands of individual source-gate cells. The shape of the source area varies among manufacturers. International Rectifier uses a hexagonal shape, and On Semiconductor and Philips use a square shape.

Figure 7-10 shows current flow in the MOSFET. As the gate voltage increases, holes are pushed away from the P+ area of the source. As the voltage increases further, a thin layer of electrons forms along the underside of the gate oxide. Current flows from the N− region of the drain along the bottom of the oxide and through the P+ and N+ areas of the source to the source metal. The shape

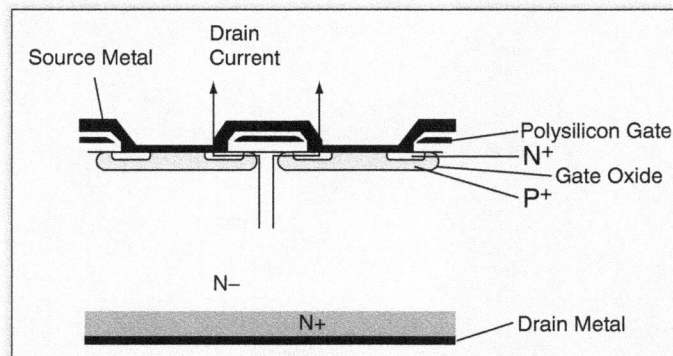

Figure 7-10: Cross section of a vertical MOSFET

of the current flow is the reason that Motorola (now On Semiconductor) called their devices TMOS.

Figure 7-11 shows representative characteristic curves for a power MOSFET. The top plot shows the constant current operation. The bottom plot shows the saturation region where current is constrained by the on resistance. The constant current region is constrained by the amount of electrons enhanced into the channel. Changing the drain-source voltage cannot increase current flow. The area on the left of each plot represents the ohmic region. In this region, r_{DSON} of the device controls current flow.

r_{DSON} is one of the most important characteristics of a MOSFET when used as a switch. In most applications, the conduction losses are larger than switching losses. Each region of the chip in Figure 7-10 contributes to on resistance. The N+ layer and metal attachment for the drain only contribute significant resistance in low voltage devices where there is a small N− epitaxial layer. The N− epitaxial layer is the major component of on resistance in devices rated above 100 V. On resistance is directly proportional to voltage rating with all other parameters equal. In low voltage devices, the resistance of the channel and the region between cells is also a major contributor to on resistance. The P+ region is sandwiched between two N regions and forms a parasitic JFET that also contributes to on resistance. Manufacturers have optimized the gate-source region to the point that on resistance below 0.01 Ω is very common in low voltage devices.

The gate voltage to fully turn on a MOSFET is greater than the source voltage, by as much as 10 V for a standard drive device. In low resistance MOSFETs, this means that the gate voltage is several volts above the drain voltage when operating in the ohmic region. In a buck converter design, this requires a bootstrap or other supply to provide the higher voltage to drive the gate. An alternative for buck converter designs is to use a P channel device. This allows the drive to pull the gate to the negative rail, as shown in Figure 7-12. The drive circuit must be designed to limit the gate-source voltage to a safe level for the MOSFET. The epitaxial layer and drain connection layer are P material in a P channel device, so the current flow is due to minority carriers. This makes the devices much slower with larger on resistance compared to N channel

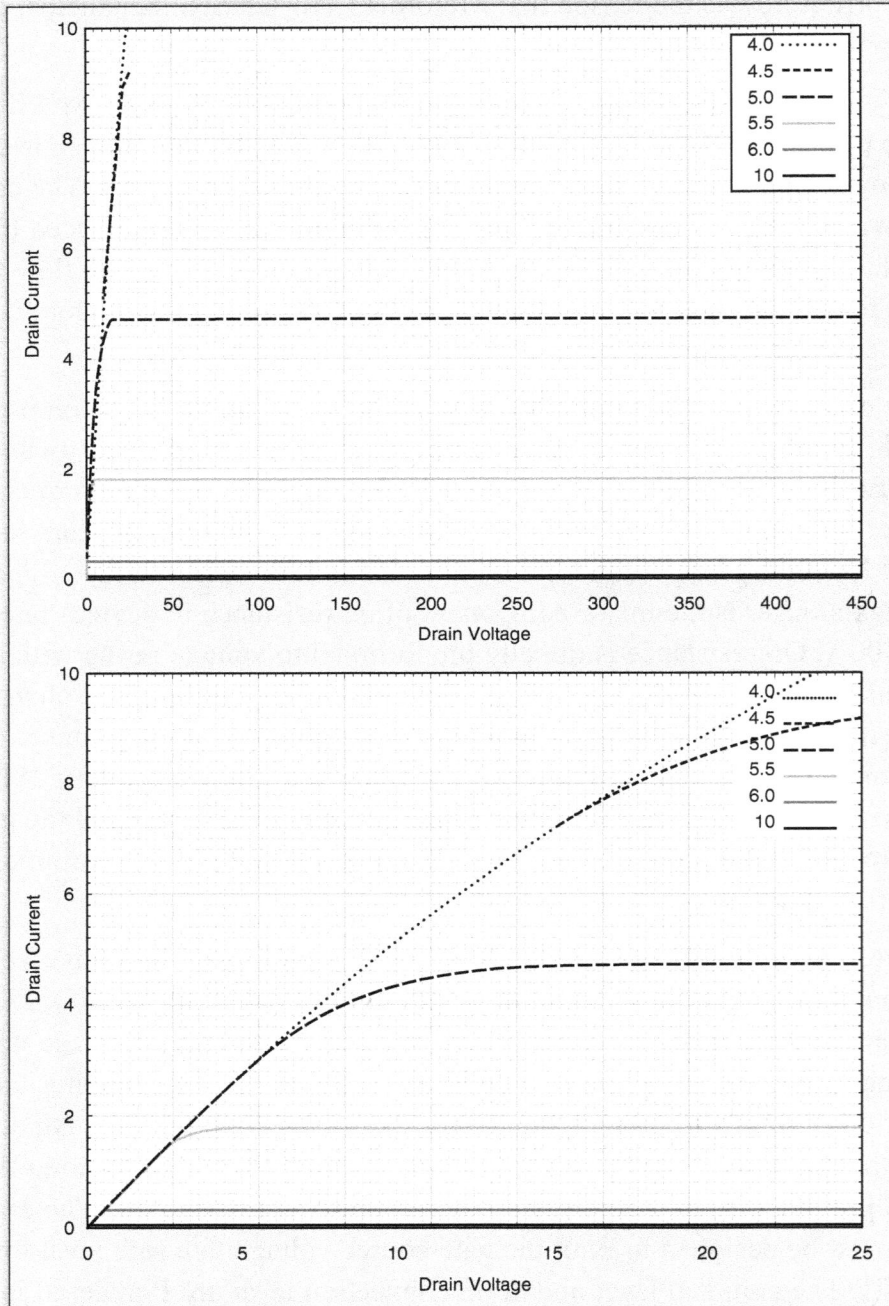

Figure 7-11: Representative characteristic curves for a power MOSFET

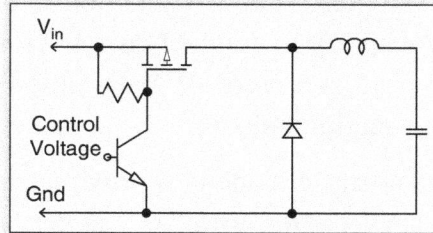

Figure 7-12: A P channel device allows the drive to pull the gate to the negative rail

devices. P channel devices have significantly larger chip areas for the same current rating as an N channel device. This limits P channel devices primarily to buck converter designs. There are significantly fewer P channel part numbers available because of the small number of applications.

The thickness of the gate oxide between the gate and the source material controls the threshold voltage and the gate-source breakdown voltage. Standard drive MOSFETs typically have ± 20 V gate breakdown voltage with threshold voltage around 4 V. Low voltage drive MOSFETs typically have ± 20 V gate breakdown voltage with threshold voltage around 2 V. Logic level MOSFETs are designed to be driven directly by either TTL or CMOS logic with ± 12 V gate breakdown voltage and threshold voltage around 1 V.

It is extremely important to ensure that the gate voltage does not exceed the maximum value listed in the data sheet. The oxide in a standard drive MOSFET is on the order of 80 nm. It can be as small as 50 nm in a logic level MOSFET. Voltage above the maximum gate-source voltage can easily puncture the gate oxide, which leads to immediate permanent damage to the MOSFET.

MOSFETs contain a PN junction between the source and drain that is an actual diode. The diode conducts current in the opposite direction of "normal" current flow from drain to source. This intrinsic diode has the same area for current flow as the MOSFET, so the current rating is the same as the MOSFET. It also has the same voltage rating as the MOSFET.

Power MOSFETs will conduct current from source to drain when turned on just as well as they conduct from drain to source. The current conduction in the MOSFET is due to electron flow from the source through the enhancement channel into a N− region, so there is no PN junction to oppose current flow. The intrinsic diode and the ability to conduct current from source to drain allow the MOSFET to be used as a rectifier. We will look at this use in the Synchronous Rectification section below.

The forward voltage of the intrinsic diode is approximately the same as for other epitaxial diodes and can vary from 0.5 V to 2.0 V at very high currents. The forward voltage decreases with increasing temperature. Minority carriers have increased mobility at higher temperatures, so the on voltage of the diode decreases. This effect is the opposite of the on resistance for the MOSFET, which increases with increasing temperature. Reverse recovery time is approximately the same as fast rectifiers—on the order of 100 ns. Reverse recovery time is generally larger for higher current MOSFETs. Reverse recovery time can preclude using the intrinsic diode in higher frequency circuits. In that case, a fast discrete diode can be used in parallel with the MOSFET to ensure that the external diode controls reverse recovery time. Totem-pole circuits such as off-line bridge circuits are problematic when using the intrinsic diode because the MOSFETs can turn on significantly faster than the intrinsic diode can turn off. This presents the possibility for current shoot-through in a totem-pole arrangement. International Rectifier, Ixys, and Advanced Power Technology manufacture devices that include a FRED in the same package as a MOSFET to ensure that reverse recovery is controlled by an ultra-fast diode rather than the slower intrinsic diode.

An advantage of MOSFETs is that the on resistance has a positive temperature coefficient. This allows devices to be used in parallel for higher current or lower total on resistance. If a device in parallel should begin to draw more current, it will heat up and self-limit. Bipolar devices have a negative temperature coefficient for saturation voltage, so they are more prone to thermal runaway (especially if transistors are used in parallel). Parasitic oscillation in the gate circuit is a common problem with using MOSFETs in parallel. Advanced Power Technology has an excellent application note regarding using MOSFETs in parallel.

Gate Drive

MOSFETs are significantly easier to drive than bipolar transistors because they are voltage devices rather than current devices. The power necessary to turn on a MOSFET is that which is necessary to charge the gate to the operating voltage (typically 10 V for standard drive devices). At first it would seem that it is only necessary to supply enough charge to bring the gate-source capacitance to 10 V. However, the high voltage on the drain interacts with the voltage on the gate through the gate-drain capacitance. This interaction is called the Miller effect.

Figure 7-13 shows a simplified model of the MOSFET as a switch and three capacitors. There is no charge on the gate-source capacitance when the gate is held at ground by the gate resistor. The gate-drain capacitor is charged to the full 350 V and holds 22 nC. We model S1 and S2 closing at the same time. When the switches close, the input voltage will need to supply the 22 nC to discharge the gate-drain capacitor as well as the 15 nC to charge the gate capacitance. The actual charge needed for the gate-drain capacitor is much larger than the 22 nC calculated as a first approximation because the gate-drain capacitance is voltage-dependent. It is 63 pF with 350 V on the drain, but it increases to 2500 pF at 0 V drain voltage. Device manufacturers provide the total amount of gate charge necessary to completely turn on the device because the voltage dependency complicates any analytical determination.

The Miller effect also causes an unexpected gate voltage waveform when turning the device on and off. Figure 7-14 shows the drain and gate voltage turn-on waveforms when the gate is driven by a constant voltage through a low value resistor and the drain load is a constant current source. The resistance of R2 combined with the high input voltage creates a nearly constant current drive of 1 A. The load is a 4 A constant current, which is very much like an inductor in

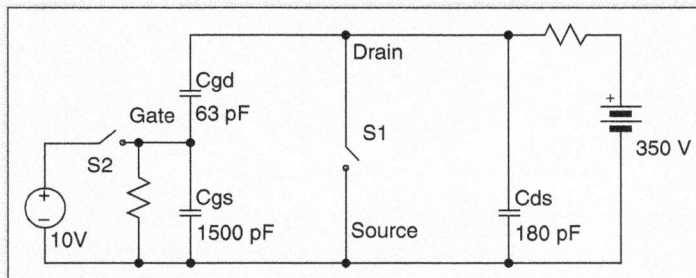

Figure 7-13: Simplified model of the MOSFET as a switch and three capacitors

continuous mode operation or like a transformer primary driving a continuous mode inductor. During time t_1, the constant current creates a linear ramp of gate voltage. Interval t_2 begins when the gate threshold voltage is reached. The device begins drawing drain current, which causes drain voltage to fall. The drain voltage falls at a fast rate, so the charge necessary to discharge the gate-drain capacitor consumes most of the input current. The drain voltage is zero at the end of interval t_3, so the current begins to charge the gate capacitance to 10 V, creating a second ramp of voltage. The model in Figure 7-14 is very close to the operation of a real forward converter or boost regulator driven by a current limited output from a control IC. The interval t_2 will vary depending on the drain current. If the current is higher than the 4 A in the example, interval t_2 will be longer. Discontinuous operation will result in t_2 that is almost zero. The

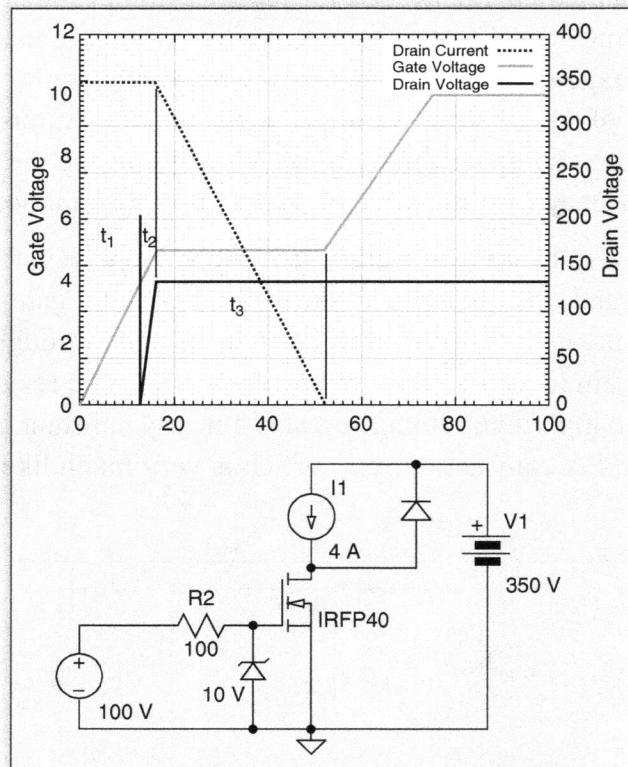

Figure 7-14: Drain and gate voltage turn-on waveforms when the gate is driven by a constant current and the drain load is a constant current source

gate voltage when t_2 begins is determined by the transconductance of the device. The drain characteristic curves as shown in Figure 7-11 allow determining the gate voltage for a given drain current.

Figure 7-15 shows how the process is reversed when turning the device off. In Figure 7-14, the current waveform has a very slight tilt during t_3 that does not show up in the plot because of the very high transconductance of the MOSFET. The same is true during t_1 of Figure 7-15.

Figure 7-16 shows the capacitance graph for the IRFPF40 from the data sheet on the International Rectifier website. The graph shows how C_{iss} changes with drain voltage. The C_{iss} measurement is taken with the drain and source shorted for AC, so C_{iss} is the sum of the input capacitance and the gate-drain capacitance (C_{rss}), which are in parallel for the test measurement. The gate-source capacitance is composed of the capacitance between the gate polysilicon and the source metalization as well as the capacitance between the gate polysilicon and the source P– and N material. The gate-source capacitance is independent of voltage. The data sheet table gives the capacitance values at 25 V. This particular data sheet shows that manufacturers are not always careful with the data

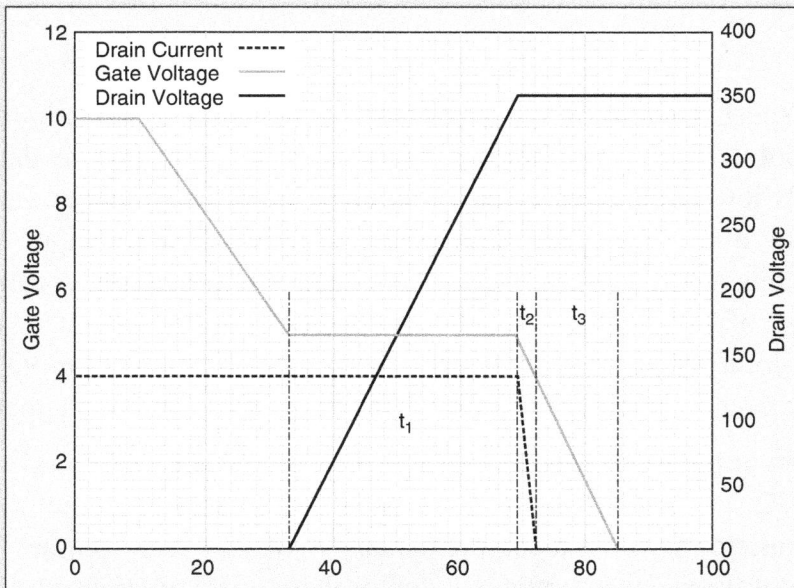

Figure 7-15: Reversal of the process shown in Figure 7-14

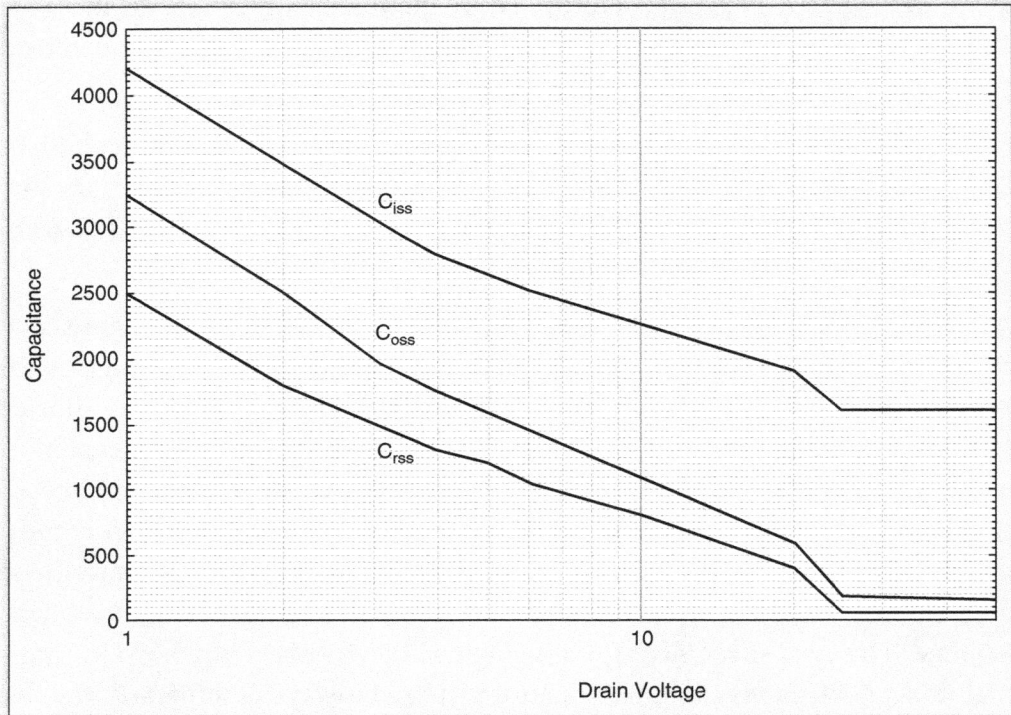

Figure 7-16: Capacitance graph for the IRFPF40

in the data sheet. The discontinuous change in capacitance shown in the graph at 25 V is not likely in a real device. Indeed, earlier versions of the data sheet have slightly lower capacitances at 1 V drain voltage. The dramatic rise of C_{iss} between 10 V and 1 V V_{DS} accounts for the majority of the total gate charge of 120 nC for this device. The gate-drain capacitance is essentially constant above drain voltage of 25 V, so the total gate charge does not change much with changes in drain voltage. International Rectifier Application Notes 937, 944, and 947 are excellent sources of more detailed MOSFET drive characteristics.

Non-isolated designs and single switch circuits are usually driven directly from the control IC. The gate drive circuit is usually a totem-pole output, a complementary transistor drive, or a complementary MOSFET drive. Figure 7-17 shows each of these configurations. Each of these configurations provides low impedance and high current drive. The rise and fall times of the MOSFET

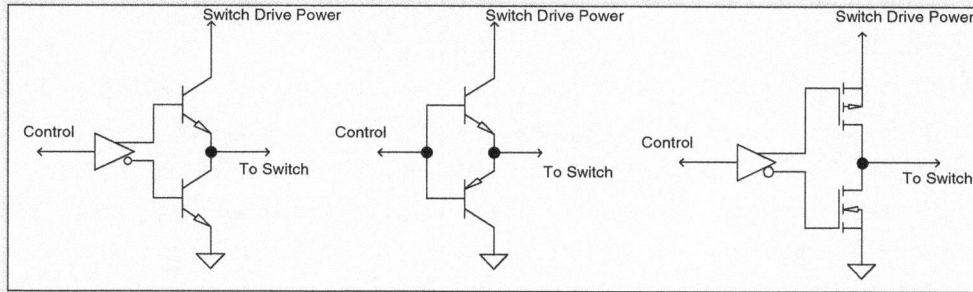

Figure 7-17: Totem-pole output, complementary transistor drive, and a complementary MOSFET gate drive circuit

switch voltage are governed entirely by how fast the gate drive circuit can move the gate charge. Lower impedance drives will switch faster and reduce switching losses. Each of the drive circuits provides fast switching by providing a saturated connection to the rails. Notice that the totem-pole and MOSFET drive circuits require a buffer that provides complementary switching. This arrangement is necessary to ensure that there is no shoot-through current as the control voltage transitions between on and off. The emitter follower complementary bipolar driver does not have that problem because the opposite transistor is always biased off by two diode drops.

The average current to drive a MOSFET gate is fairly small because it only flows during the time you charge or discharge the gate. However, the peak current can be quite large. Using our example IRFPF40 MOSFET at 100 kHz with switching time of 100 ns, we would require a peak current of 1.2 A at turn-on and 1.2 A at turn-off. The average current is only 24 mA. As we saw in Chapters 4 and 5, it is sometimes necessary to add a series resistance between the control IC and the MOSFET gate to reduce the peak current to a level compatible with the control IC. This will increase switching losses and slow switching time. If the peak gate current is too large for the control IC, you can maintain switching time by using an external emitter follower or complementary MOSFET circuit, as shown in Figure 7-17.

Bridge circuits and two switch forward and flyback designs require a floating circuit to drive the high side MOSFET. Figure 7-18 shows a high side driver IC that is designed for off-line supply operation. The driver IC uses level shifting

FETs to drive an internal latch. The IC provides signal conditioning to ensure that the high and low side drive remain synchronized. Additionally, they include circuitry to prevent accidental high side turn-on due to noise. If the high side should turn on while the low side device is on in a half bridge circuit, the MOSFETs will be destroyed immediately due to shoot-through current from the positive rail to the negative rail. The IR2110 and similar devices use a bootstrap technique that we saw in Chapter 4. The bootstrap capacitor C1 charges up when the high side switch is off and the low side device is on. The capacitor supplies a voltage referenced to the source and is independent of the positive rail. The high side drive voltage from C1 voltage is large enough to fully turn on the high side MOSFET. The bootstrap capacitor must be large enough to run the high side circuit inside the IR2110 and also provide sufficient charge to turn on the MOSFET. The level shifting circuit and high side drive work up to 600 V.

Figure 7-19 shows high side drive using a pulse transformer. This circuit is best suited for relatively constant duty cycle applications. If the duty cycle is too large, there is danger of saturating the transformer and the secondary voltage will go to zero. Saturation due to duty cycle is less of a problem for switching frequencies above 100 kHz. Resistors R1 and R2 guarantee that the transformer continues to supply current once the gate is fully charged. R2 is sized to allow a fairly constant charging current for the gate. R1 ensures that the gate voltage

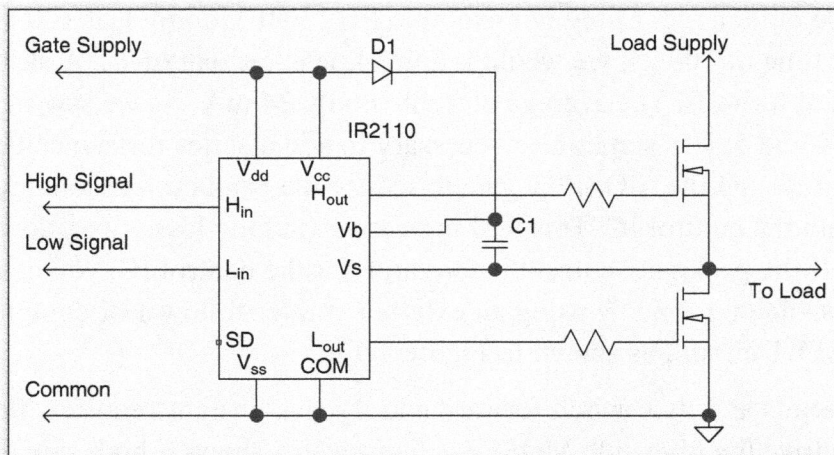

Figure 7-18: High side driver IC designed for off-line supply operation

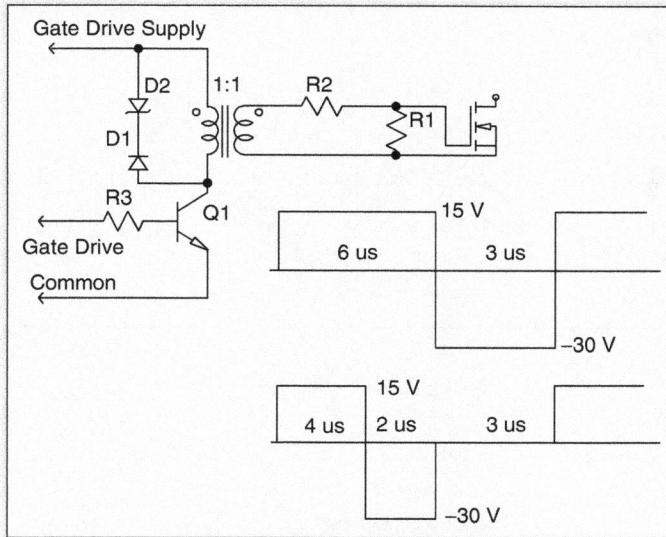

Figure 7-19: High side drive using a pulse transformer showing constant volt-second operation

stays below the gate breakdown voltage and it discharges the gate charge when the drive is removed. The zener diode across the primary is necessary to ensure enough voltage to dissipate all of the magnetizing inductance current at the highest duty cycle. The zener voltage must be low enough that the gate voltage stays below the breakdown voltage. All transformers require that the volt-seconds when charging the MOSFET gate are equal to the volt-seconds when discharging the gate.

The waveforms in Figure 7-19 illustrate the equal volt-seconds requirement. The waveforms show the transformer primary voltage for 67% duty cycle and 44% duty cycle. At 67% duty cycle, the volt-seconds provided by D2 are just enough to completely dissipate the magnetizing current. The combination of the R1/R2 voltage divider, gate drive supply voltage, and zener voltage must be balanced to ensure adequate drive at low duty cycle and to avoid excessive gate voltage at high duty cycle. Q1 must withstand at least 45 V in this design. It is likely that the duty cycle may be limited to a value less than 67% to allow R1 enough time to discharge the gate capacitance.

Figure 7-20 shows an improved circuit that drives current during both charge and discharge of the gate. The capacitor is required to ensure that magnetizing

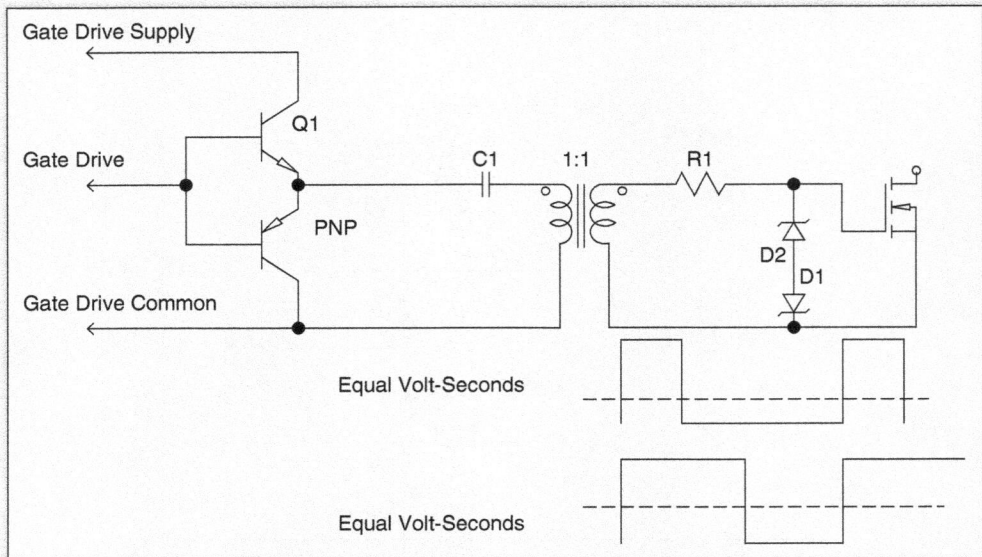

Figure 7-20: Improved circuit that drives current during both charge and discharge of the gate

current averages to zero. The AC waveform across the transformer will have equal volt-seconds, as shown in the waveforms in Figure 7-20. This has two consequences. The first is that the turn-on voltage increases as duty cycle decreases. Setting adequate turn-on voltage at high duty cycle will result in a voltage that could exceed the gate-source breakdown at low duty cycle. The second consequence is that turn-on and turn-off times are asymmetrical. The back-to-back zener diodes in Figure 7-20 limit the gate voltage to allow a larger drive voltage and a large duty cycle range.

Figures 7-19 and 7-20 showed only the high side MOSFET driven by a transformer. Figure 7-21 shows a further improved circuit that drives both high and low side MOSFETs from the same transformer. This circuit drives the MOSFETs in a more symmetrical fashion. The gate voltage on the "off" MOSFET is driven negative while the "on" gate is driven positive. This improves turn-off and gives more margin to ensure that both MOSFETs cannot conduct at the same time.

So far, we have looked at drive circuits aimed at creating the shortest possible switching time in order to reduce the switching losses. Increased RFI is a con-

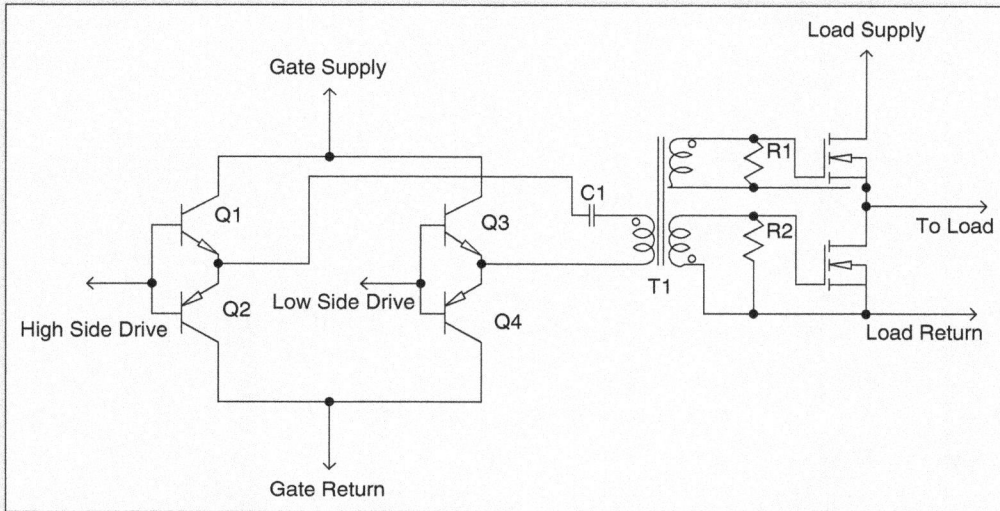

Figure 7-21: A circuit that drives both high and low side MOSFETs from the same transformer

sequence of short switching times. We can modify the gate drive to decrease RFI at the expense of more switching loss in the drain circuit. There are three effects that determine the amount of noise. The first is the switching frequency and its associated duty cycle. The second is the rise times and fall times of the drain voltage and current. The third is ringing due to resonances, particularly in the output rectifier circuit.

A perfect rectangular wave will include harmonics that have a $\sin(x)/x$ amplitude pattern. A perfect 50% duty cycle rectangular waveform has $\sin(x)/x$ zeros that coincide with the even harmonics. Other duty cycles will have varying amounts of harmonics. The switching transition time changes the relationship of harmonic energy from a single $\sin(x)/x$ to a ($\sin(x)/x * \sin(y)/y$) function. It is possible to adjust the rise time to interleave the zeros of each $\sin(x)/x$ function to reduce the overall level of harmonics. There is a point of diminishing returns where increasing rise time increases power loss with no measurable decrease in harmonics. Figure 7-22 illustrates the $\sin(x)/x$ effect for both rectangular and trapezoidal pulses. These plots show the height of each of the harmonics from 1 to 7 for a rectangular (0% rise/fall) and a trapezoid (10% rise/fall) with 40% duty cycle. Notice that the longer rise time for the trapezoid has moved the energy from higher harmonics into the first six harmonics.

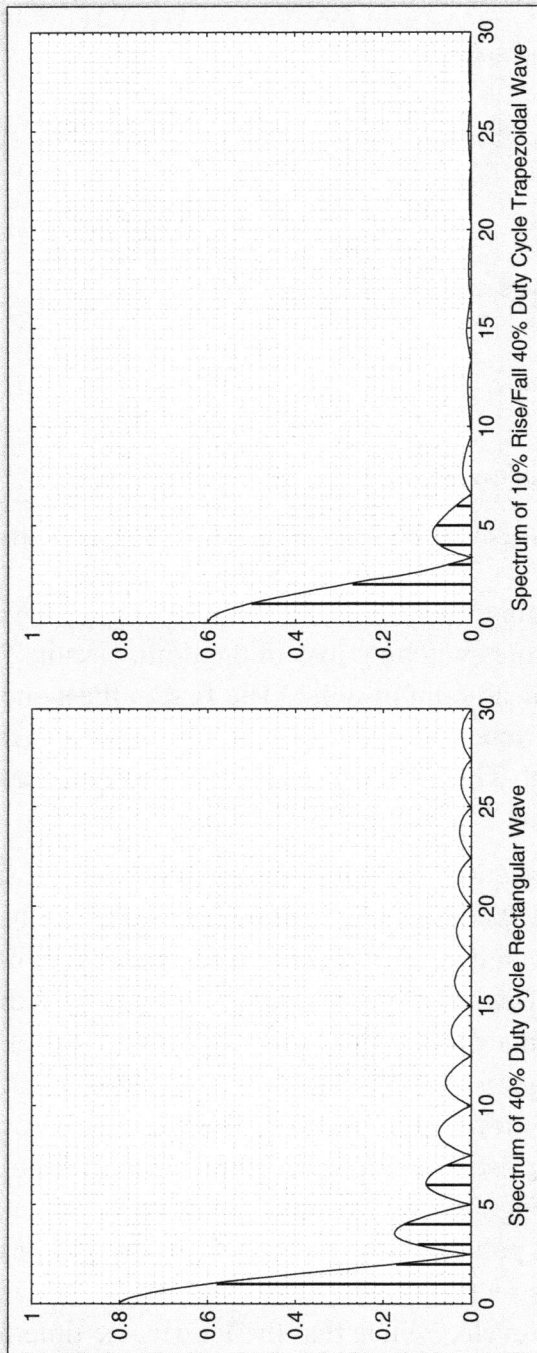

Figure 7-22: Sin(x)/x effect for both rectangular and trapezoidal pulses

Control of rise time and duty cycle will not appreciably affect the amount of energy at the switching frequency. Some control ICs are designed to minimize the energy at the switching frequency by using spread spectrum control. These ICs randomly change the oscillator frequency to spread the energy across a wider spectrum. This also spreads the energy of the harmonics. If the energy of the switching frequency is spread across 10 kHz, then the tenth harmonic energy is spread across 100 kHz. This method can produce dramatic decreases in energy, on the order of tens of dB. It is also possible to simply vary the frequency linearly with a sine wave or a triangle wave to produce an FM signal (just like an FM radio transmitter). This will also spread the energy but will not give as much reduction in energy. A triangle or sawtooth wave will give better spreading than a sine wave because of the harmonics in the modulating waveform. Linear Technology produces the LTC6902. This part is an oscillator with inherent spread spectrum capability. The frequency is set by one resistor and the amount of spreading is set by another resistor. It has four 90-degree outputs to drive four phase CPU power supplies. Setting the spreading to 20% decreases the fundamental energy by 20 dB.

Ringing due to leakage inductances and resonances in the output circuit create RF energy at frequencies that are not related to the switching frequency. These can be controlled by using RC snubbers, as described in Chapter 5. Controlling rise and fall time also has an affect on these frequencies because they are excited by the fast transitions. Increasing rise time or fall time also reaches a point of diminishing returns in reducing this source of RFI.

Safe Operating Area and Avalanche Rating

The safe operating area for a MOSFET is constrained by the current rating on the left, the power dissipation in the center, and the breakdown voltage on the right of the SOA plot, as shown in Figure 7-23. MOSFETs do not have a secondary breakdown effect as bipolar transistors do. MOSFETs have a positive temperature coefficient for resistance. If excessive current begins to flow in a region of the chip, the resistance increases and current flow shifts to other regions of the chip. The negative temperature coefficient in bipolar transistors leads to avalanche and secondary breakdown. Bipolar transistors such as the

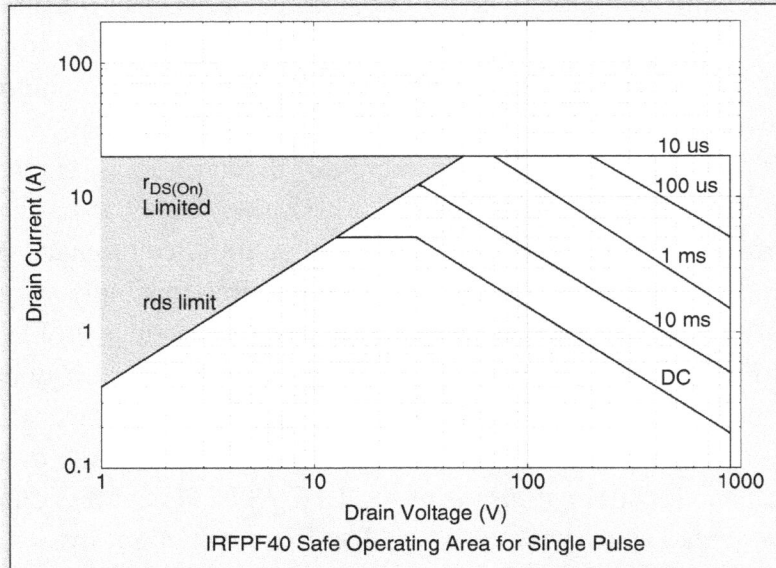

Figure 7-23: Safe operating area (SOA) plot for the IRFPF40 MOSFET

BUT11 have breakdown voltage as high as 1500 V, where the highest voltage MOSFETs handle is 1200 V. This would seem to indicate that bipolar transistors have an advantage in high voltage applications. This is true if you take adequate care to ensure that the transistor is always operated away from the secondary breakdown region. Bipolar transistors typically must be derated to ensure safe operation, so there is typically little advantage in high speed switching circuits. Figure 7-23 shows the SOA plot for the IRFPF40 MOSFET.

The area in the upper left of the MOSFET SOA plot is unreachable because $r_{DS(ON)}$ will not allow operation in that area. The boundaries of the safe operating area are determined solely by the energy necessary to raise the junction temperature to the maximum value. There are two classes of operating temperature for MOSFETs: 150C° and 175C°. The SOA plot in the data sheet assumes that the case is held at 25C°. The DC operating line is the appropriate limit for switching power supply operation. The safe operating area will also decrease as the case temperature increases. The data sheet will give the appropriate derating information for elevated temperature operation.

The SOA plot shows very large currents for very small pulse widths. The peak current limit is set by the current capability of the bond wires. The lower current limit at DC is constrained primarily by the bonding pads and the source metallization as well as the temperature of the junctions. As long as adequate heat sinking is provided, the RMS value of the current can be quite large.

Modern MOSFETs for switchmode power supplies are typically rated for use up to the drain-source breakdown voltage where avalanche current begins. Rugged devices are characterized for maximum avalanche current and repetitive avalanche energy. Using the avalanche rating of a MOSFET allows a design without using snubbers or clamps to limit the drain voltage when the MOSFET is turned off. International Rectifier Application Note 1005 gives design information to take advantage of the avalanche rating of a MOSFET.

Avalanche failures in a MOSFET occur because the parasitic bipolar transistor turns on. Figure 7-24 shows the cross section of a power MOSFET. The N and P regions of the source combine with the N− epitaxial layer to create a parasitic NPN bipolar transistor. If current through the P region becomes too large, the voltage drop across the inherent resistance of the large P region can increase to the point that it forward biases the PN source junction. The portion of peak current flow is at the corner of the P region, as shown in Figure 7-24. The long

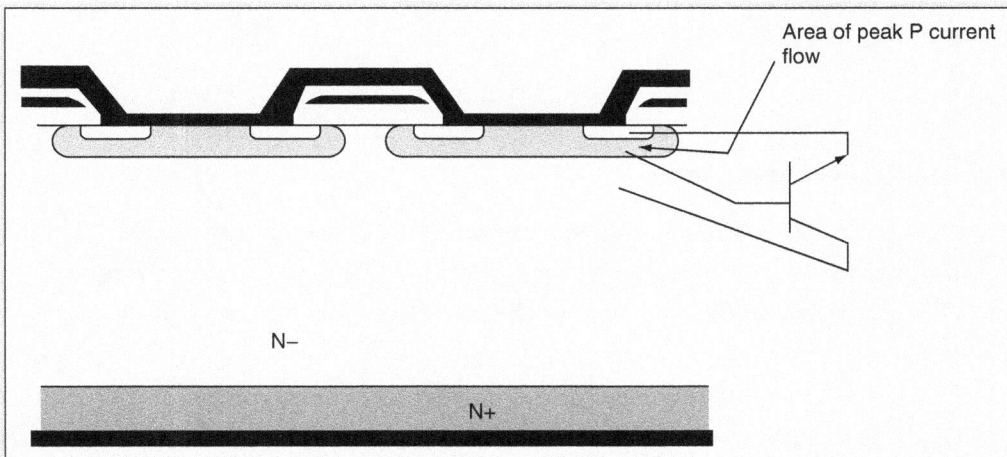

Figure 7-24: Cross section of a power MOSFET showing a parasitic NPN transistor that causes avalanche failures

travel path allows the voltage drop. The voltage drop turns on the bipolar transistor, which will cause the device to fail. The failure occurs because the NPN transistor has a negative temperature coefficient, so turning it on under high current high temperature conditions will cause the affected cell to hog most of the drain current. Rugged devices use heavily doped P regions to reduce the resistance to the point that the transistor cannot be turned on except under excessive avalanche currents.

The parasitic bipolar transistor can also be turned on by rapid increase in drain-source voltage during the time that the body-drain diode is recovering from conduction. The rapid rise in drain voltage while the minority carriers are recombining can forward bias the emitter-base junction and turn on the transistor, leading to device failure. Rugged MOSFETs for switching supply use have a *dv/dt* rating for reverse recovery time of the intrinsic diode. There are two ways to handle the *dv/dt* problem. The first is to design the turn-on of the MOSFET to ensure that the *dv/dt* rating is met. The other solution is to use one or two external fast recovery diodes in conjunction with the MOSFET switch.

A single diode can be used in parallel with the MOSFET to keep the intrinsic diode off. A diode in series with the MOSFET will block intrinsic diode current flow. Figure 7-25 shows these approaches. It is only important to consider the diode *dv/dt* rating in cases where the MOSFET is turned on during the time of diode reverse recovery. D1 must have forward voltage below the forward voltage of the MOSFET diode. D2 ensures that the MOSFET diode can never turn on.

Synchronous Rectification

Computer processor voltages are decreasing and current levels are increasing. It is common for processor cores to run at only 1.25 V and consume tens of amps. 3.3 V (instead of 5.0 V) is now the most common voltage for logic circuits. These lower voltages mean that diode conduction losses can be a significant percentage of the output power. Consider a 15 V class silicon Schottky diode that has 0.36 V forward voltage. When this diode is used in a buck regulator with 5 V input and 3.3 V output at 15 A, the diode dissipation will be 1.84 W when conducting and as much as 0.33 W while off because of the 100 mA

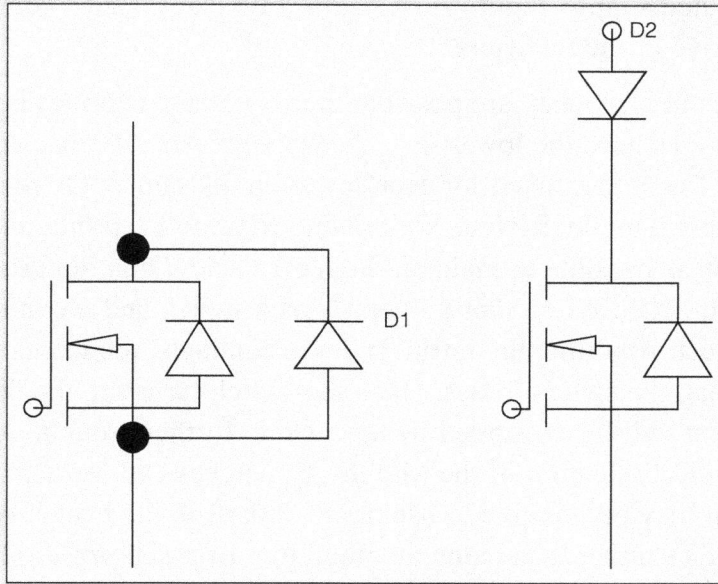

Figure 7-25: A single diode can be used in parallel with the MOSFET to keep the intrinsic diode off. A diode in series with the MOSFET will block intrinsic diode current flow

of reverse current. A forward converter will use two diodes and the total loss in the diodes will be on the order of 5.9 W for both diodes. This means 11% of the power is lost just in the diodes.

Many of the non-isolated circuits we looked at in Chapter 4 used two MOSFETs as switches. These circuits require that the lower MOSFET is turned on after the top MOSFET has completely turned off. The nonoverlapping drive requires that the circuit use the intrinsic diode of the lower MOSFET. The time delay between diode turn-on and MOSFET turn-on is very small, so losses in the intrinsic diode are quite small. Once the lower MOSFET is fully turned on, it shorts the intrinsic diode and all the current flows through the MOSFET. There is some loss due to the reverse recovery of the diode. Additionally, the MOSFET turn-on must not exceed the *dv/dt* rating of the diode.

This control of the MOSFET switch mimics the action of a diode and is called synchronous rectification because the switch action is synchronous with the switching control. Synchronous rectification can also be applied to the secondary side of an isolated supply. Figure 7-26 shows a typical forward converter with transformer drive of the rectifiers.

Great increases in efficiency are possible in low voltage supplies because low voltage MOSFETs have the lowest $r_{DS(ON)}$. Even greater efficiencies are possible if MOSFETs are paralleled for even lower on resistance. On resistance has a positive temperature coefficient, so it is important to keep the junction temperature as low as possible to maintain high efficiency. It is important to design the drive to the MOSFET so that it operates as a switch and the intrinsic diode does not conduct. Any time the intrinsic diode conducts, the losses will be much larger than the switch losses. Once the switch turns on, the diode reverse recovery current will be consumed by the switch, further reducing efficiency. The lower power dissipation of the MOSFET switches can further improve the supply, since it may be possible to use the PC board as the heat sink for surface mount MOSFETs rather than using an aluminum heat sink for diode rectifiers.

The circuit in Figure 7-26 has the problem of multiple windings and control of the gate drive to the MOSFETs. Since synchronous rectification is only practical for 5 V or lower outputs, we can take advantage of the low voltage of the transformer winding. For outputs up to 5 V, the voltage range will remain below the 20 V rating of the gate-source, even in a universal input supply. Figure 7-27 shows a circuit that uses the voltage of the secondary power winding to supply the gate control voltage for the rectifiers. Note that the gate drive

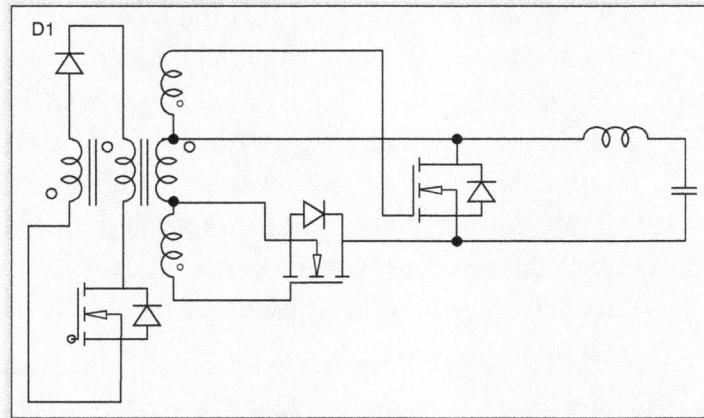

Figure 7-26: Typical forward converter with transformer drive of the synchronous rectifiers

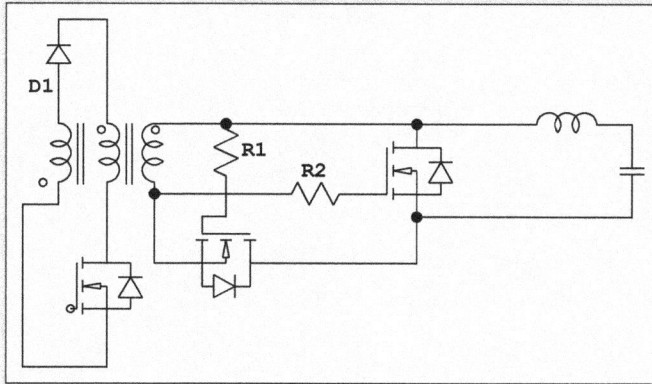

Figure 7-27: Circuit that uses the voltage of the secondary power winding to supply the gate control voltage for the rectifiers

of the freewheeling MOSFET is guaranteed by the voltage across the main rectifier MOSFET intrinsic diode.

Notice that we use the main MOSFET switch in the "negative" lead of the transformer. Usually, a diode is placed in the "positive" lead of the transformer and the "negative" lead of the transformer is connected to circuit common. Both methods have identical results. Also notice that the intrinsic diode of the main MOSFET is always reverse biased. The intrinsic diode in the freewheeling switch will conduct unless the MOSFET is turned on quickly.

There are numerous problems with ensuring adequate drive to the MOSFET rectifiers, especially in wide input range off-line supplies. Many control IC manufacturers are producing ICs for the secondary side of off-line supplies to facilitate high efficiency synchronous rectification. International Rectifier has the IR1176 that handles control of MOSFETs, as in Figure 7-28. This IC is designed for outputs of 5 V or lower. The MOSFETs are set up so that the intrinsic diodes will conduct unless they are shorted by the MOSFET. The power supply is bootstrapped by using the two intrinsic diodes and D1 and D2 as a full wave bridge. Once V_{DD} reaches 5 V, the IC starts and drives the switches. The IR1176 uses internal VCOs and dead time control for each MOSFET. The control logic takes advantage of the property that the duty cycle will change rather slowly as the power supply tracks changes in load and input.

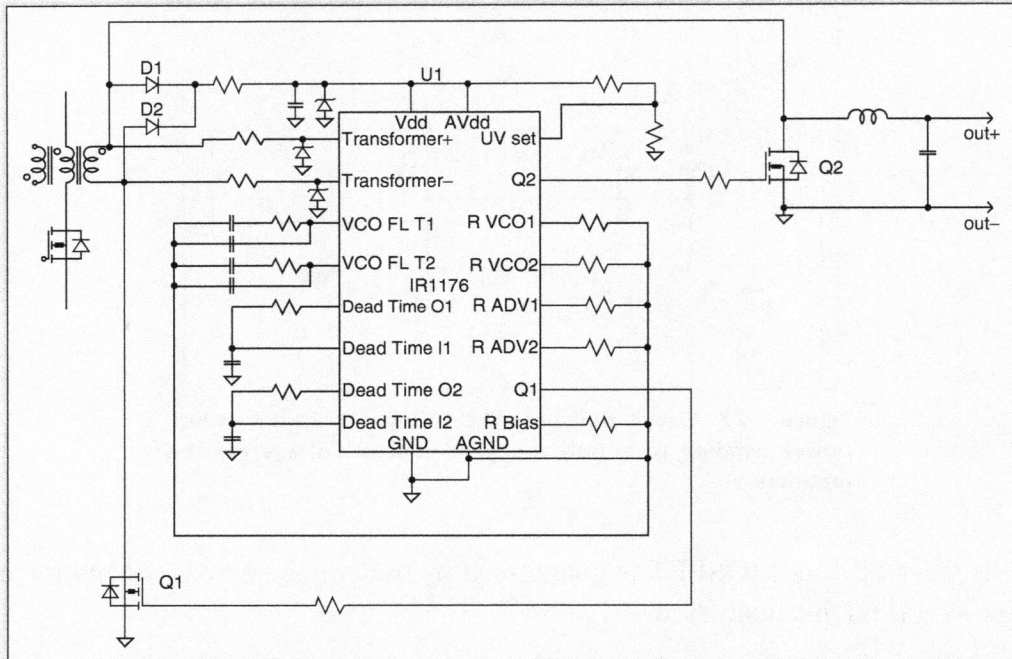

Figure 7-28: Synchronous rectification using the IR1176

The VCOs and the dead adjustments use the rising and falling edges of the transformer signal to track the changes in duty cycle with a phase locked loop.

Linear Technology has the LT3900 and LT3901 that use switching information from the primary side to provide adequate control to the MOSFET rectifiers. Figure 7-29 shows a representative circuit for the LTC3900. As in the case of the IR1176, the intrinsic diodes of the MOSFETs conduct until the IC begins to drive the MOSFETs. R1, D1, and C2 delay the drive to the main switch to give the sync information time to operate the rectifier switches. The LTC3900 contains a timer to ensure that drive to the MOSFETs is removed if the sync signal quits. It also monitors the current through Q3 so that drive can be removed if the inductor current starts to go negative.

Figure 7-30 shows a representative circuit for the LTC3901 that is designed for full wave drive circuits. It is interesting that the drive requires only two

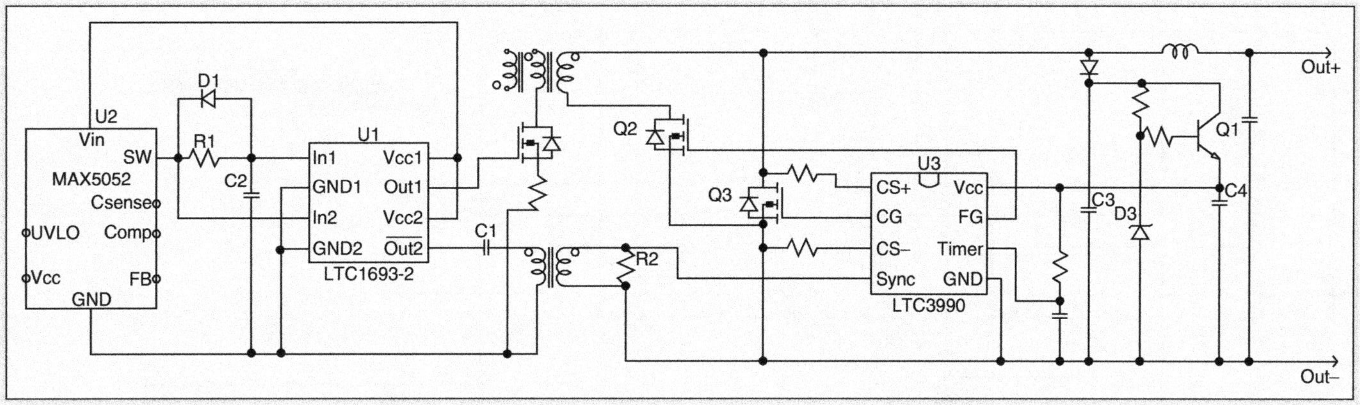

Figure 7-29: Synchronous rectification using the LTC3990

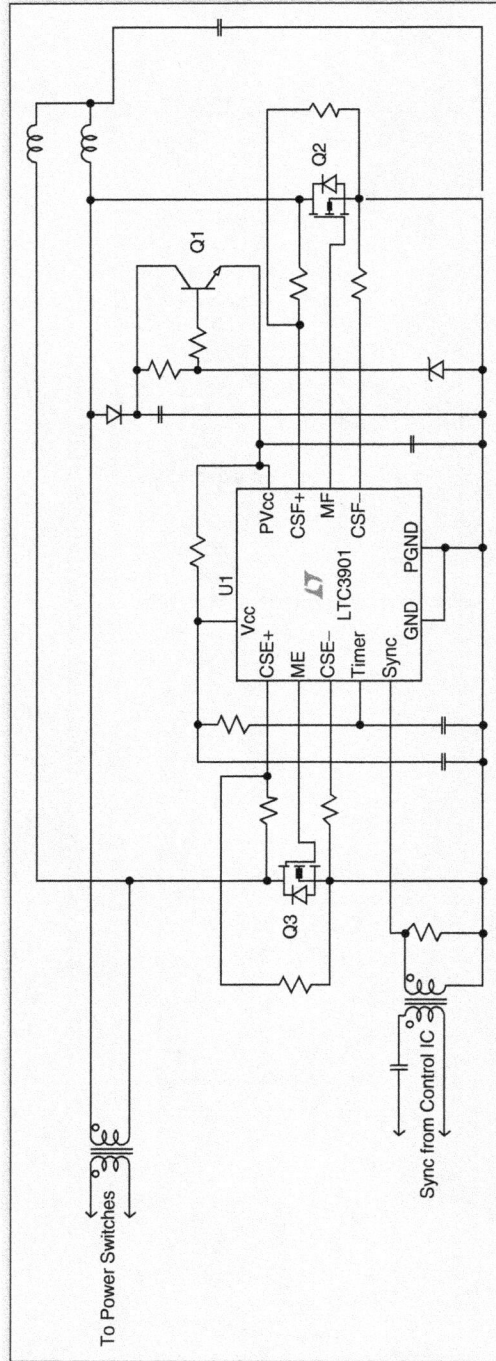

Figure 7-30: Representative circuit for the LTC3901 that is designed for full wave drive circuits

228

switches without a center tap to provide full wave rectification. This IC has the same timer and current sense circuits as the LTC3900.

Sense FETs

A power MOSFET is actually many thousands of MOSFETs in parallel. Each small transistor has a fairly large resistance and so contributes a small amount to total drain current. When current mode control first became popular, MOSFET manufacturers realized that they could isolate a small number of cells on a MOSFET source and bring out these sense cells separately from the main source. The ratio of the current in the sense cells is a relatively constant ratio of the total current in the device. These devices allow sensing of device current without the need for a high power, low resistance current shunt. A sense FET is a five-terminal device. A Kelvin terminal and sense terminal are added to the drain, source, and gate terminals. These devices are available, but the number of devices from each manufacturer is quite small.

Package Options

MOSFETs come in a large variety of packages. The number of package options continues to grow as MOSFETs become the preferred device for switching applications. Each package is aimed at a particular use. Metal can packages such as the TO-3 power transistor package are no longer generally available. The highest power devices are TO-247 flat packages that have the same pin spacing as the TO-3 metal package but with significantly less volume. TO-247 packages typically handle many hundreds of watts. TO-220 devices typically handle powers from 10 W to perhaps 150 W. Both TO-247 and TO-220 packages are designed for through-hole mounting of the leads and mounting of the heat dissipating surface to a large metal heat sink. Both the TO-220 and TO-247 packages are available in isolated and non-isolated styles. The isolated styles provide creepage and clearance distances for safety agency requirements when the heat sink is at chassis potential. International Rectifier supplies devices in packages they call FULL-PAK that completely enclose the TO-220 or TO-247 device in plastic, so an insulating mount like mica or Silpad is not necessary. These devices trade slightly higher thermal resistance for greatly

229

reduced mounting cost. Another new package now offered by manufacturers is a TO-220 or TO-247 size package with no mounting hole. The package is intended to be mounted to the heat sink with a spring clip directly over the chip. This produces a much more uniform mounting to the heat sink for improved thermal contact.

There are a large number of surface mount packages available for lower power applications. The most interesting is the D-PAK package, which is designed to be soldered to a large copper heat sink area on a printed circuit board. Some of these devices have power dissipation ratings to 50 W. Another interesting surface mount package is available from International Rectifier. The DirectFET package uses a thin metal can for the drain connection that exposes a metal surface to the ambient air. The source and gate connections are exposed on the die underneath the drain can for soldering directly to the printed circuit board. This design gives large heat dissipation with a very low profile. Another interesting package is the BGA package available from Fairchild Semiconductor. This package is similar to the International Rectifier DirectFET package in that it has an integral heat sink exposed to ambient with source and gate connections under the package.

IGBT Devices

Insulated gate bipolar transistors (IGBTs) are semiconductor devices that are a hybrid of a MOSFET and a bipolar transistor. Figure 7-31 shows the cross section of two types of IGBTs. These devices use a vertical structure, as we have seen in previous power devices. An IGBT is essentially a PNP bipolar transistor where the base current is supplied by a parasitic FET between the collector and base. Figure 7-32 shows an equivalent circuit illustrating the parasitic elements inside an IGBT. The structure is essentially the same as an N channel MOSFET with an added P+ layer at the back. This additional PN junction adds a series diode to the structure that eliminates the intrinsic diode of the MOSFET.

MOSFETs have fairly large $r_{DS(ON)}$ when the voltage rating goes above 500 V. This greatly increases the conduction losses compared to bipolar transistors of the same voltage rating. Additionally, MOSFET conduction losses increase as temperature increases due to increasing on resistance.

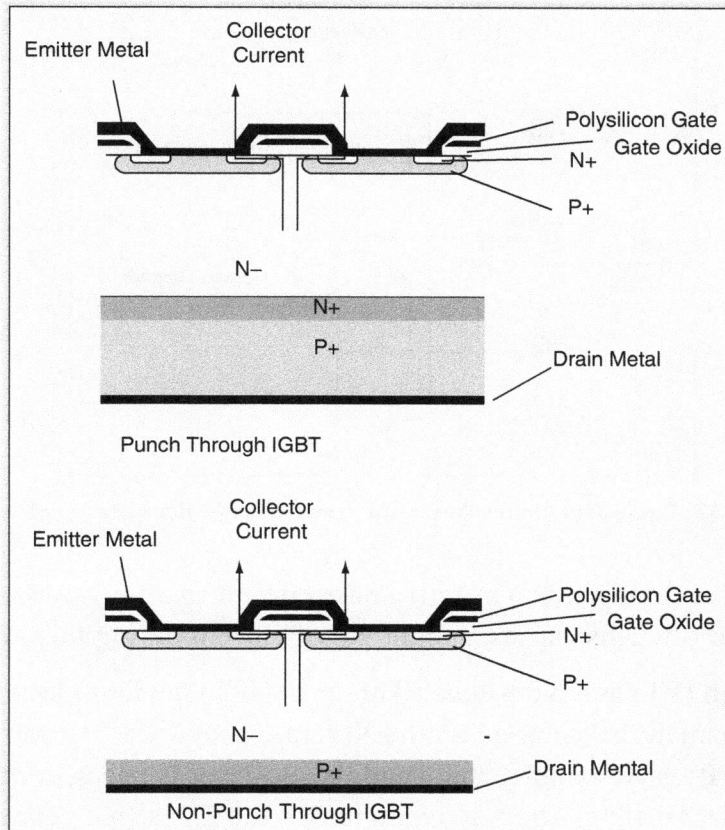

Figure 7-31: Cross section of two types of insulated gate bipolar transistors (IGBTs)

The P+ layer in an IGBT injects minority carriers into the N− epitaxial layer, which improves the conductivity in the N− drift region. This is the same effect that occurs in bipolar transistors. The conductivity modulation from the P+ layer makes the on voltage relatively constant across all voltage ratings.

The PNP transistor in the IGBT does not fully saturate, so the on voltage never goes below one diode drop. Typical on voltage is 1.0–3.0 V. Turn-off time is much better in an IGBT than a bipolar because there is no storage time due to saturation. Electron flow in the IGBT ceases immediately when gate voltage is removed, but the current in the drift region due to holes continues until all the holes recombine. The base junction of the PNP has no external connection, so

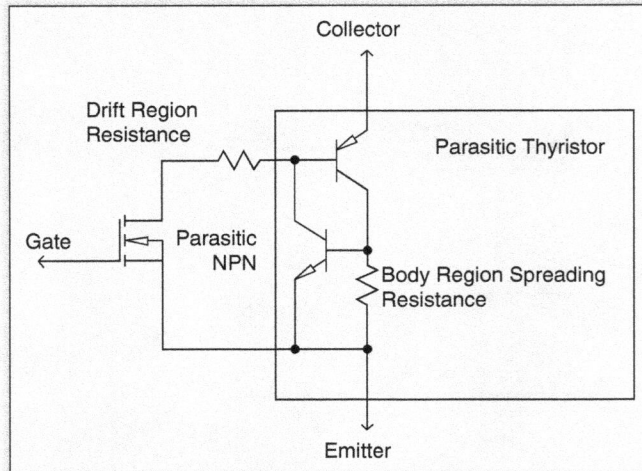

Figure 7-32: Equivalent circuit illustrating the parasitic elements inside an IGBT

it is not possible to generate a negative base current to draw out minority carriers during turn-off. This causes a slight tail of current during turn-off.

Punch Through (PT) and Non-Punch Through (NPT) devices have different current tail characteristics based on the structure above the P+ layer. In a PT transistor, the P+ layer is fairly thick and the N+ layer is necessary to control minority carrier lifetime. The N+ region makes the PT device very much like a MOSFET with an added P+ layer. The large minority carrier region causes PT devices to have a negative temperature coefficient for on voltage. The minority carrier lifetime control in the N+ layer increases speed at the cost of higher on voltage.

The NPT device is manufactured on a much thinner die than a PT device. The P+ region is very lightly doped so the number of minority carriers injected into the drift layer is much smaller. The lower number of minority carriers decreases turn-off time. The drift region resistance dominates the temperature coefficient, so the temperature coefficient of on voltage is positive in NPT devices. The thinner die limits the voltage rating to around 600 V. NPT devices are the usual choice for a switchmode power supply.

MOSFETs are most cost effective when the voltage is lower than 250 V at all switching frequencies. IGBTs are most cost effective for voltages above 1000 V and frequencies up to a few hundred kilohertz. In the voltage range between 250 V and 1000 V, the IGBT is the clear choice below 20 kHz and the MOS-FET is the clear choice above ~ 150 kHz. Choosing an IGBT or MOSFET for a design in the crossover region requires careful analysis of cost and efficiency.

An IGBT behaves much like a MOSFET, except for on voltage and turn-off current. Turn-on time is controlled by the characteristics of the internal MOS-FET and is quite fast. The gate drive characteristics are also dominated by the input and Miller capacitances. The gate drive voltage has the same stepped shape we saw with MOSFETs due to the Miller effect. The gate drive must use total gate charge rather than the capacitances to accurately reflect circuit operation. All of the circuits we looked at for MOSFETs will work for IGBTs. IGBTs also do not suffer from secondary breakdown. The safe operating area is bounded by maximum current and maximum voltage, as in the MOSFET. IGBTs are also rated for avalanche energy for use near the maximum voltage.

CHAPTER 8

Inductor Selection

- Properties of Real Inductors
- Core Properties
- Designing a Powder Toroid Choke Core
- Choosing a Boost Converter Core

Inductor Selection

The basic circuits in earlier chapters assumed ideal inductors. Real inductors have the properties of inductance, resistance, and capacitance. Losses due to the core material are the equivalent of an additional resistance. The primary properties we must consider when choosing a core are permeability, magnetic losses (hysteresis), maximum flux density (saturation), and core temperature.

Properties of Real Inductors

One of the most important properties of a real inductor is flux leakage. Any flux that is not contained inside the center of the inductor is a potential source of EMI. Inductors are either shielded or unshielded. An unshielded inductor can have a magnetic core or can be air-wound. Air-wound inductors will only be practical at very high frequencies because of the small amounts of inductance that can be obtained in a reasonable size. Most practical unshielded inductors will require a magnetic core. The core will reduce the amount of leakage flux by concentrating the majority of the flux in the core and immediately outside the wire, but there will still be significant flux outside the core. Figure 8-1 shows two unshielded magnetic core shapes. The bobbin shape can be partially shielded by using a magnetic sleeve that covers the outside of the coil but does not close the magnetic loop. Both of these shapes are typically used in unshielded surface mount power inductors.

Any shielded inductor will need a magnetic core that completely encloses the wire of the inductor. Figure 8-2 shows core geometries that provide varying degrees of shielding. The bobbin shape can also be fully shielded by arranging

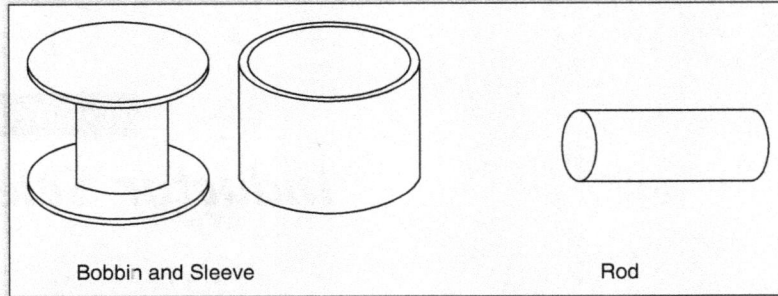

Figure 8-1: Unshielded magnetic core shapes: bobbin and sleeve, and rod

the sleeve to cover both ends of the bobbin. Many surface mount shielded inductors use this configuration.

All inductors have parasitic capacitance caused by the proximity of adjacent turns. Inductors usually have negligible capacitance at switching power supply frequencies because the core usually allows a small number of turns.

The resistance of the wire of the inductor becomes important when the DC current through an inductor becomes quite large. An inductor for a 10 A output

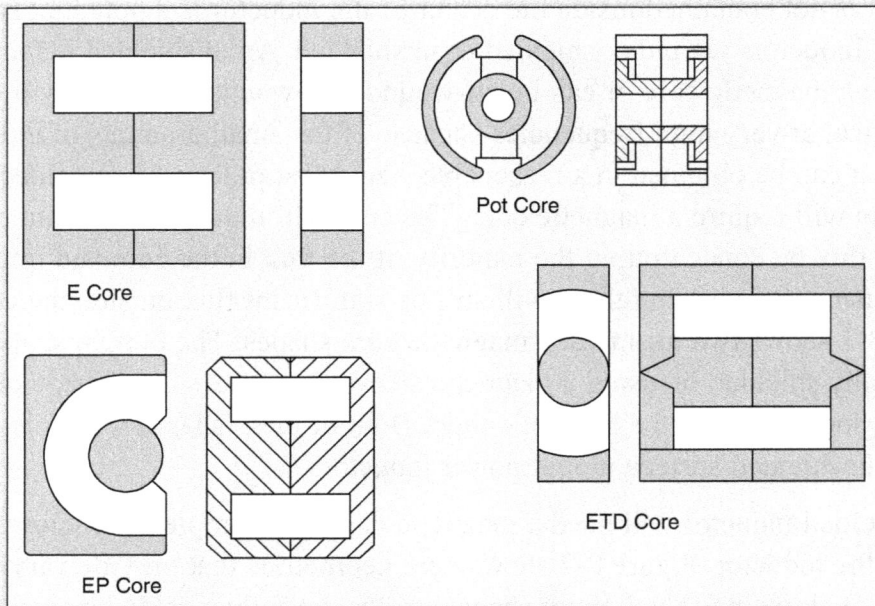

Figure 8-2: Different core geometries for shielded inductors

238

Table 8-1 Current capacity versus AWG wire size

AWG Wire Size	200 A/cm^2 5 C° rise	400 A/cm^2 15 C° rise	600 A/cm^2 30 C° rise	800 A/cm^2 45 C° rise
8	16.5	33.0	49.5	66.0
10	10.4	20.8	31.2	41.6
12	6.53	13.1	19.6	26.1
14	4.11	8.22	12.3	16.4
15	3.26	6.52	9.78	13.0
16	2.58	5.16	7.74	10.3
17	2.05	4.10	6.15	8.20
18	1.62	3.25	4.88	6.50
20	1.02	2.05	3.08	4.10

supply may use 60 cm of wire for the 10 to 20 turns required. If we choose #20 wire, it will only have 0.02 Ω of resistance, which seems quite small. However, that 0.02 Ω will generate 2 W of heat (Table 8-1). The total surface area of 60 cm of #20 wire will be only 15 cm^2, so the temperature rise will be quite large. A second problem is that the 0.02 Ω and 10 A current will create a 0.2 V voltage drop across the inductor. In most DC supplies, that voltage will be significant. Table 8-1 shows a sample of current capacity versus wire size versus current density. The temperature rise in Table 8-1 is for single layer coils. Temperature rise will be greater for multiple layer coils.

Some core manufacturers, such as Micrometals, give temperature information versus wire size for various powdered iron cores in their catalog. Others are not as helpful.

Standard wire tables usually give a single value of current capacity based on 300 A/cm^2 for coil and transformer service or 600 A/cm^2 for house wiring service. This single value will give a starting point for laboratory measurements. None of the wire manufacturers gives very useful information on current density versus current capacity versus temperature rise versus wire size. The best practice is to wind a target core and measure the temperature rise in the laboratory while passing the target DC through the winding.

The wire also contributes resistance due to the AC flowing in the coil because of skin effect. The resistance of a conductor increases as frequency increases. There is a cutoff frequency where skin effect begins in a round conductor and can be calculated from:

$$f = 124/d^2 \qquad\qquad (8\text{-}1)$$

where f is in MHz and d is in mils.

Resistance due to skin effect above the cutoff frequency can be reduced by using multiple wires in parallel, by using Litz wire, or by using flat conductors. Multiple wires in parallel will reduce the amount of resistance caused by skin effect, but the extra turns increase the amount of capacitance between the windings. Litz wire is composed of multiple conductors that are woven in a way that minimizes capacitance between the conductors. Litz wire is very effective below 500 kHz. It becomes less effective as the frequency goes above 3 MHz because of the capacitance effects between the strands. It is not generally practical to use flat conductors (ribbon or strip) for inductors because of the number of turns that are usually required. Flat conductors will also have much larger parasitic capacitance when used in an inductor.

Core Properties

Magnetic materials are composed of very small magnetic domains (on the order of several molecules). These domains are randomly oriented when no external magnetic field is applied. As an external field is applied, these domains start to align with the external field. The domains absorb some of the energy from the field as they align with the external field. As the external field strength is increased, more domains align completely with the external field. When all of the domains align with the field, any further increase in field intensity will not affect the material. This is saturation. If the external field is decreased, the domains will attempt to return to the original orientation. However, not all of the domains will return to the original orientation and some of the absorbed energy will be turned into heat instead of being returned to the external field. This property is called hysteresis. Hysteresis losses are the magnetic equivalent of dielectric losses. Both types of losses are due to the interaction between the electrons of the material and the external field.

Figure 8-3(a) shows a classic B-H curve that plots the flux in the material (B), which is a measure of domain alignment vs. the strength of the applied field (H). The shape of this curve shows the hysteresis of the material. This classic B-H curve actually shows the maximum limits of B and H. If you apply an

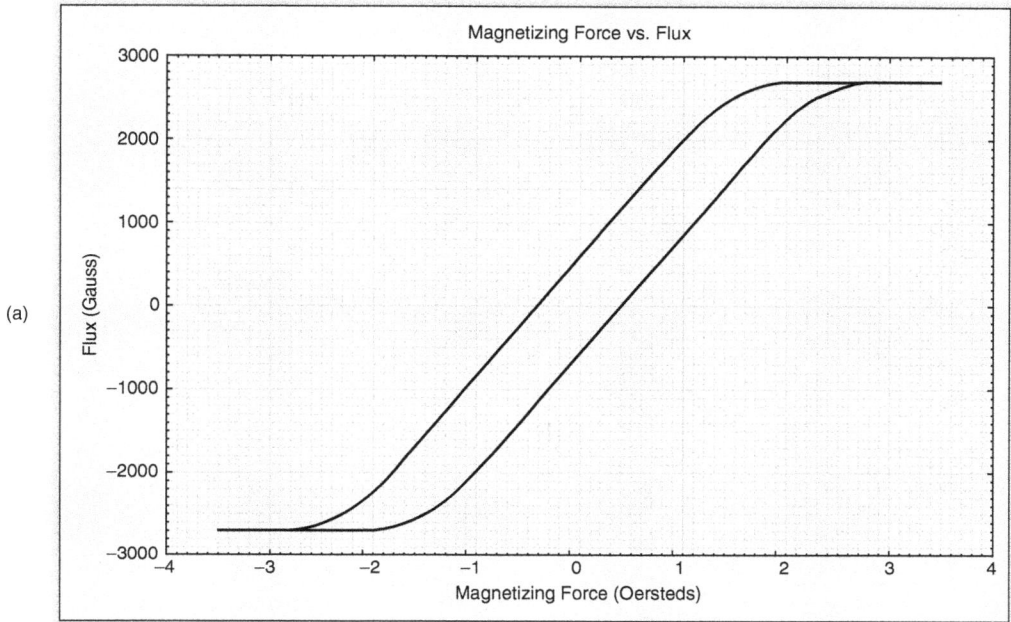

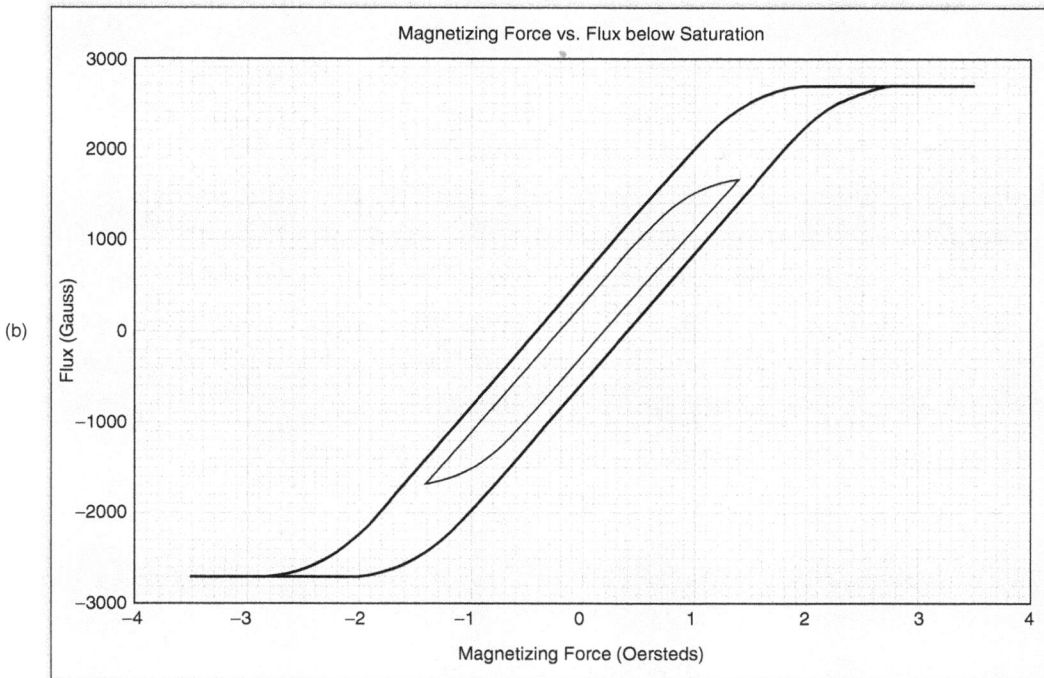

Figure 8-3: (a) Classic B-H curve that plots the flux in the material (B), which is a measure of domain alignment versus the strength of the applied field (H); (b) Curve with a smaller enclosed area produced by an alternating field that does not have enough intensity to produce saturation.

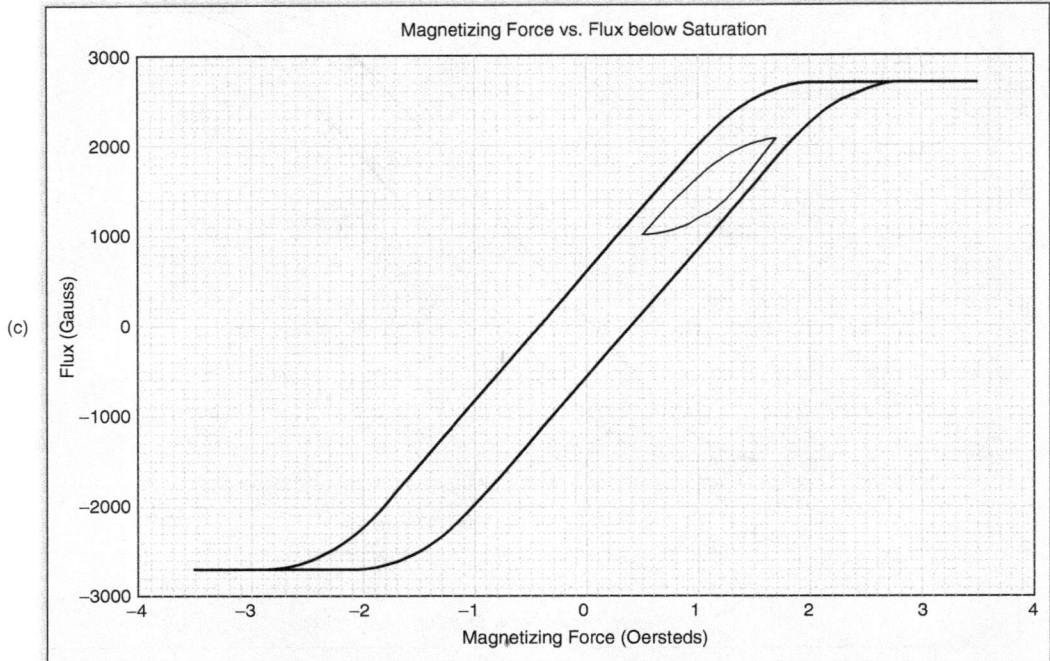

Figure 8-3: (c) Curve with an even smaller enclosed area produced by a varying field that does not change direction

alternating field that does not have enough intensity to produce saturation, you will get a curve with a smaller enclosed area, as shown in Figure 8-3(b). This is the case with a transformer in a bipolar drive circuit like push-pull or bridge. Finally, if you apply a varying field that does not change direction, you get a curve with an even smaller enclosed area, as shown in Figure 8-3(c). This would be the case for a buck converter filter choke.

The magnetic force applied to an inductor's core is given by:

$$H = \frac{0.4 \times \pi \times N \times I}{l_e} \tag{8-2}$$

where N is number of turns, I is current in amps, and l_e is the magnetic path length of the core in cm, and H is in oersteds (Oe).

Permeability is the slope of the B-H curve. Figure 8-4 shows a core being operated in two different portions of the B-H curve of a real core. Permeability is the slope of the line through the center of the operating B-H curve. The permeability in Figure 8-4(b) is much smaller than in 8-4(a). When a core saturates,

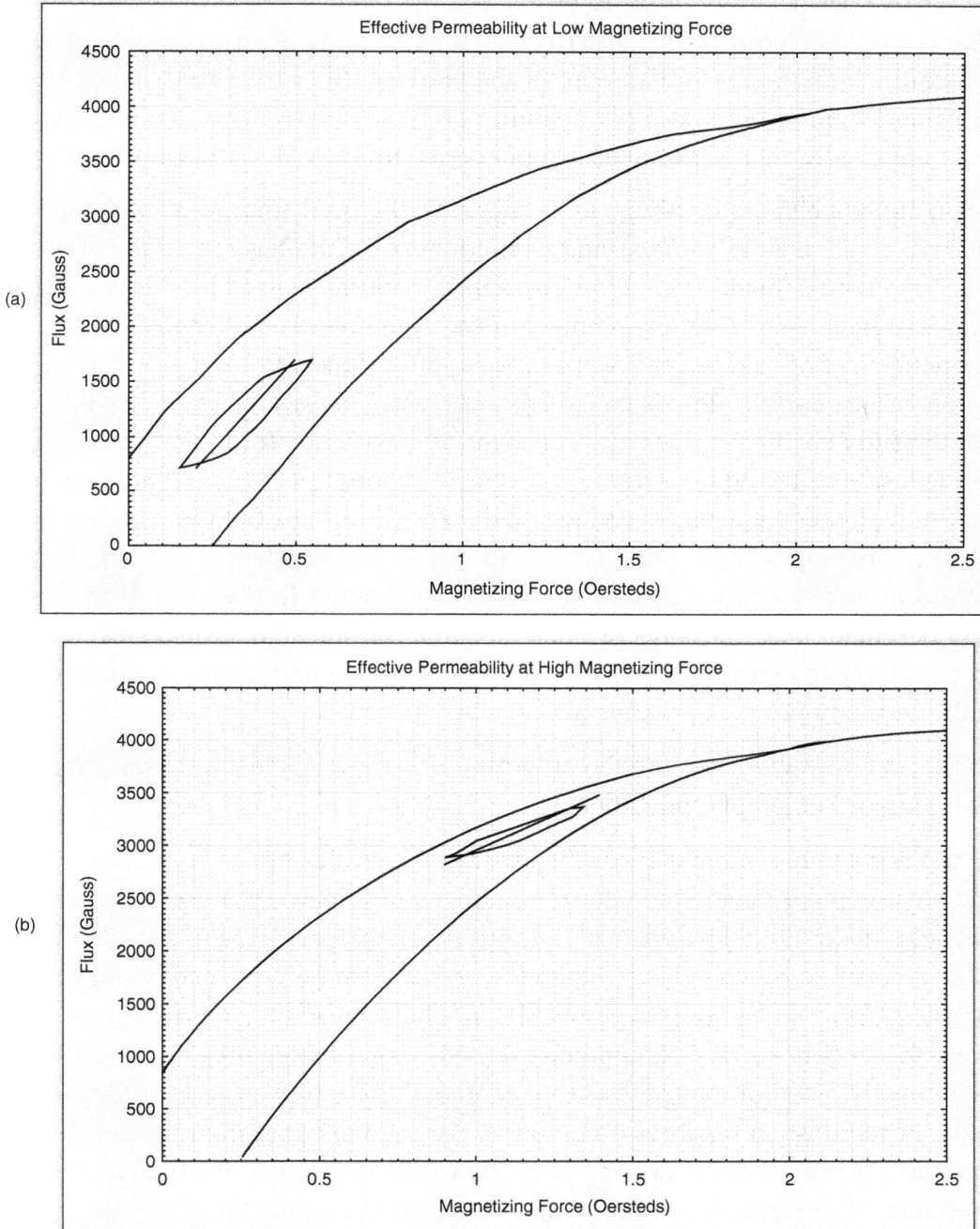

Figure 8-4: Core being operated in two different portions of the B-H curve; permeability in (b) is much smaller than in (a)

the relative permeability drops to near 1 and the inductance becomes roughly what it would be without the core (usually a very small value). The inductance of a coil is increased by the amount of the permeability of the core. An inductor using a core with relative permeability of 20 will have twice the inductance of an identically shaped inductor using a core with relative permeability of 10.

If you apply a changing voltage to an inductor, the field strength will change and will create a field with varying strength over the dimensions of the core. When the core is made from a low resistance material such as iron, this varying field will induce small loops of current flow in the material of the core. The magnetic field of this current opposes the applied field and results in heating the core. This current flow is called eddy current. The heat generated in the core by eddy currents causes an equivalent AC resistance for the inductor. The physical size of the eddy currents is inversely proportional to the frequency of the field. Higher frequencies create smaller eddy currents, but there are more of them, creating a larger total loss. Solid iron or steel is shaped into thin laminations to reduce eddy currents at power line and audio frequencies. Iron or steel is seldom used for cores of inductors above power line frequencies because the eddy current losses cannot be limited by controlling the thickness of the laminations, as is possible at power frequencies.

Ferrite and powdered iron are the only two core materials that are useful for switching power supply inductors.

Ferrite is a ceramic composed of magnetic metal oxides mixed with iron oxide. The most popular magnetic materials are manganese-zinc or nickel-zinc. The magnetic materials are mixed with an organic binder and fired in a kiln to make a ceramic. Since they are ceramics, ferrites can be produced in many shapes simply by changing the mold shape. They can also be machined after firing for smooth surfaces and precise dimensions. Ferrite cores for power use are typically made of manganese-zinc compounds for higher permeability. Eddy current loss in ferrite is quite low in the normal operating range of the material because of the insulating property of the oxides and the binder. The resistance of the material is quite high compared to metals (three to four orders of magnitude higher). Eddy current loss increases as frequency increases, but these losses are minimal compared to eddy current losses in sheet steel.

Catalogs and application notes refer to "soft ferrites." This has nothing to do with the mechanical hardness of the material, but instead refers to the magnetic B-H curve of the material. Soft ferrites have a rather low value of residual magnetism when the magnetizing force is removed, whereas hard ferrites have residual magnetism that is nearly identical to the saturation flux density. Ceramic permanent magnets are "hard ferrites."

Powder cores are made by grinding iron or other alloys into fine particles and coating them with an insulating material. Powder cores are die pressed and baked. Powder cores are usually either toroids or rods, but some manufacturers can produce other shapes such as E cores. The size of the powder limits the upper frequency of powdered cores because of eddy current losses.

Inductors for power applications are formed into a closed loop, as a general rule, in order to contain the magnetic field completely within the inductor. The amount of flux and, therefore, the energy that can be held in the core is a property of the material. The amount of flux in the closed magnetic circuit can be increased dramatically if a very small air gap is introduced into the loop. Because the relative permeability of air is 1 and the relative permeability of a magnetic material will be many thousands, the majority of the magnetic energy will be stored in the flux in the air gap. The air gap decreases the effective permeability of the core and tilts the B-H curve, as shown in Figure 8-5. Figure 8-5 represents the same material that is shown in Figure 8-4. Notice that the curve enters saturation around 25 Oe instead of 2.5 Oe. The air gap also increases the magnetizing force necessary to saturate the core because much of the applied magnetic force is stored in the air gap.

Ferrite cores can be made with a small gap machined in one of the mating surfaces of a core set to determine the amount of reduction in permeability. Powdered iron cores have an inherent distributed air gap provided by the insulating binder around each of the iron particles. This distributed air gap is a major advantage in power inductors. A core that has a discrete gap is a potential source of EMI. It is possible for the magnetic field from the gap in the inductor to interfere with other parts of the system. It is also possible for other parts of the power supply system to interfere with the field in the inductor. If a gapped core is required, it is better to use a larger core with a small gap

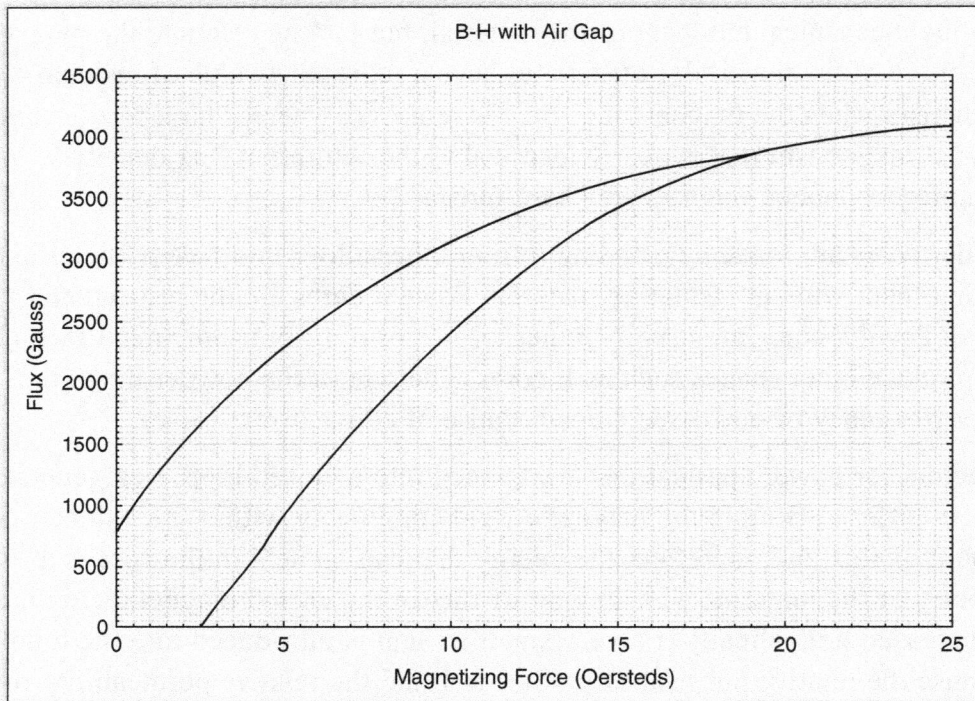

Figure 8-5: Air gap decreases the effective permeability of the core and tilts the B-H curve (core material is the same as in Fig. 8-4)

than a small core with a large gap to minimize stray fields. It is also easier to control the inductance of a core if the gap is kept small. A larger gap allows external factors like the windings of the core to interact with the magnetic field of the core.

There are several core shapes that provide a closed magnetic circuit with adequate shielding of the magnetic field. E cores with a gap in the center leg provide moderate amounts of shielding. Pot cores provide the largest amount of shielding for cores that can be easily machined to provide a gap. However, pot cores have very poor heat dissipation properties and it is hard to wind coils with large-diameter wire. The RM-, DS-, and RS-style cores are modifications to the classic pot core that allow better heat dissipation and allow larger wire sizes. These cores will still maintain adequate magnetic shielding with a gap in the center post. All manufacturers can supply these standard shapes with machined gaps in the center leg for inductor use. The final inductor can be produced from a gapped core half and an ungapped core half, or by using two gapped halves. The combinations of standard gap lengths and ungapped cores allow flexibility in the design of an inductor. A major advantage of ferrite core inductors is that the winding can be machine-wound on a plastic bobbin and the inductor is then assembled from the winding and two core halves. The core halves are normally glued together to form the final assembly. Standard gaps are usually set up to give a particular A_L value, so you will see different standard gap sizes for different materials.

Hysteresis losses in the core are proportional to both frequency and AC flux density. Figure 8-6 gives a set of power loss curves for a representative Magnetics ferrite material (Type R). The actual amount of power absorbed from

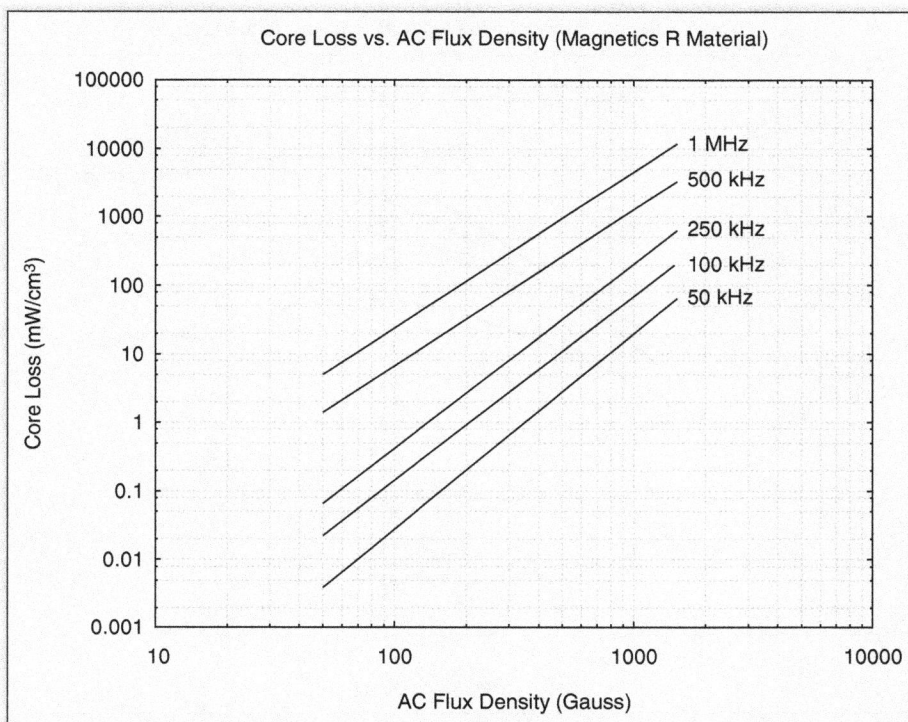

Figure 8-6: Set of power loss curves for a representative Magnetics ferrite material (Type R)

the winding and converted to heat depends on the AC flux density and the volume of the core. Figure 8-7 shows the figure of merit (B times *f*) for several Magnetics and Ferroxcube ferrite materials at 300 mW/cm³. The following formula gives the AC flux density for an inductor.

$$B = \frac{E_{AVG} \times 10^8}{4 \times A \times N \times f} \,(\text{Gauss}) = \frac{L \times \Delta I \times 10^8}{2 \times A \times N} \qquad (8\text{-}3)$$

where A = magnetic area in cm², N = number of turns, f = frequency in Hz, and E_{AVG} is the applied square wave AC voltage in volts.

Temperature affects several of the parameters of a magnetic core. Figure 8-8 shows permeability versus temperature, core loss versus temperature, and saturation flux density versus temperature for Magnetics Type R ferrite material. Permeability drops abruptly to one at a high temperature. This is the Curie tem-

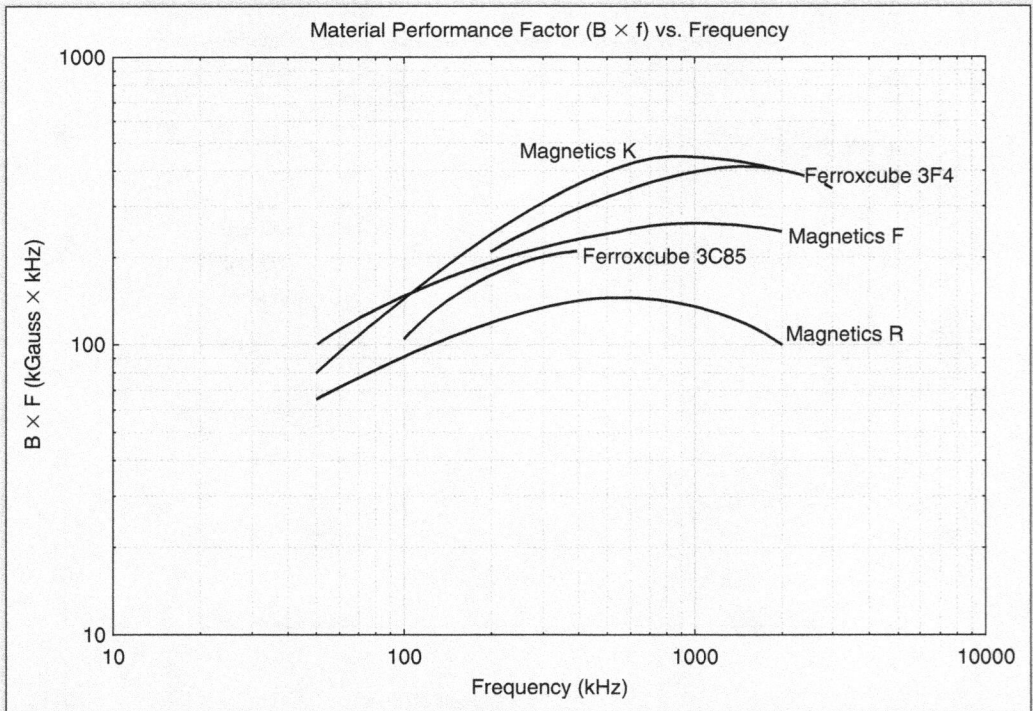

Figure 8-7: Figure of merit (B * f) for several Magnetics and Ferroxcube ferrite materials at 300 mW/cm3

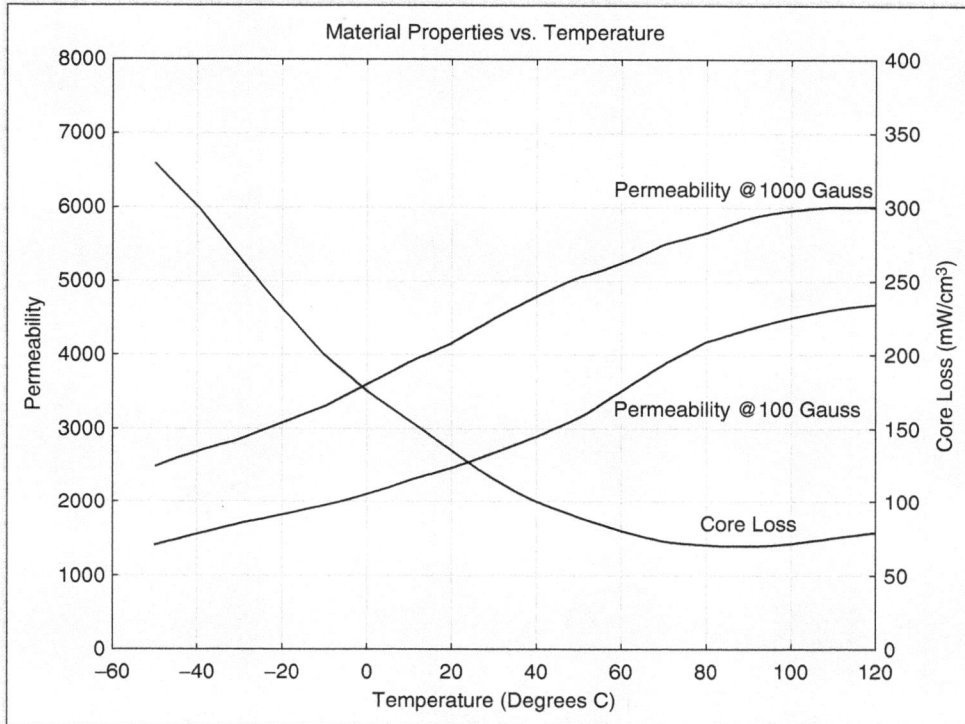

Figure 8-8: Permeability versus temperature, core loss versus temperature, and saturation flux density versus temperature for Magnetics Type R ferrite material

perature. Fortunately, for most magnetic materials, the other parts of the coil will melt before the coil gets to the Curie temperature.

DC bias current causes powder cores to have a gradual decrease in permeability as the magnetic field is increased. Figure 8-9 shows the reduction of permeability versus magnetizing force for several Micrometals materials. AC flux density has the opposite effect on permeability, as shown in Figure 8-10. For switching power supply filter chokes, the net effect tends to be no change in permeability. Figure 8-11 shows the change in permeability due to frequency. The change in inductance due to frequency will be minimal at switching power supply frequencies. Powder cores show the same type of core loss versus AC flux density as ferrites. Figure 8-12 shows core loss versus AC flux density and

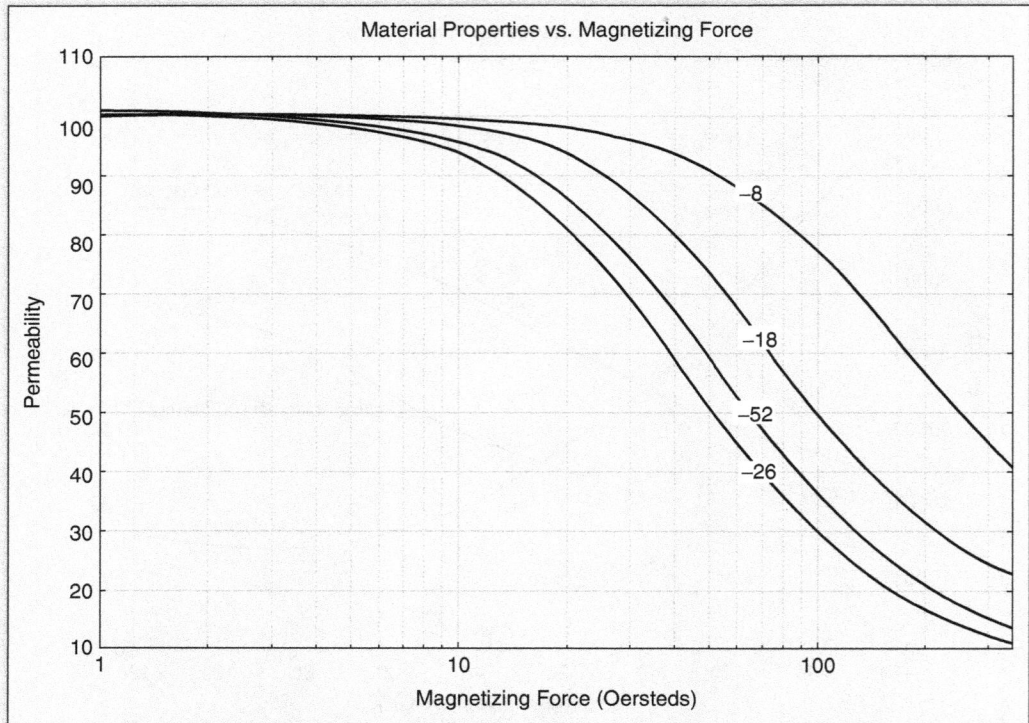

Figure 8-9: Reduction of permeability versus magnetizing force for several Micrometals materials

frequency for Micrometals −26 material. All Micrometals powder core materials show a positive temperature coefficient for permeability.

Designing a Powder Toroid Choke Core

Forward converters and buck regulators require a choke inductor to smooth out the rectangular pulses into a DC value. These supplies typically operate in continuous current mode, which means the current has a large DC component with a smaller AC component. The large DC bias current flowing through the winding will create a very large magnetic field in the core. An air gap is almost always necessary in the core to ensure that the core does not saturate. Powder core toroids are frequently chosen for smoothing chokes because of cost, control of permeability, control of EMI, and ease of assembly.

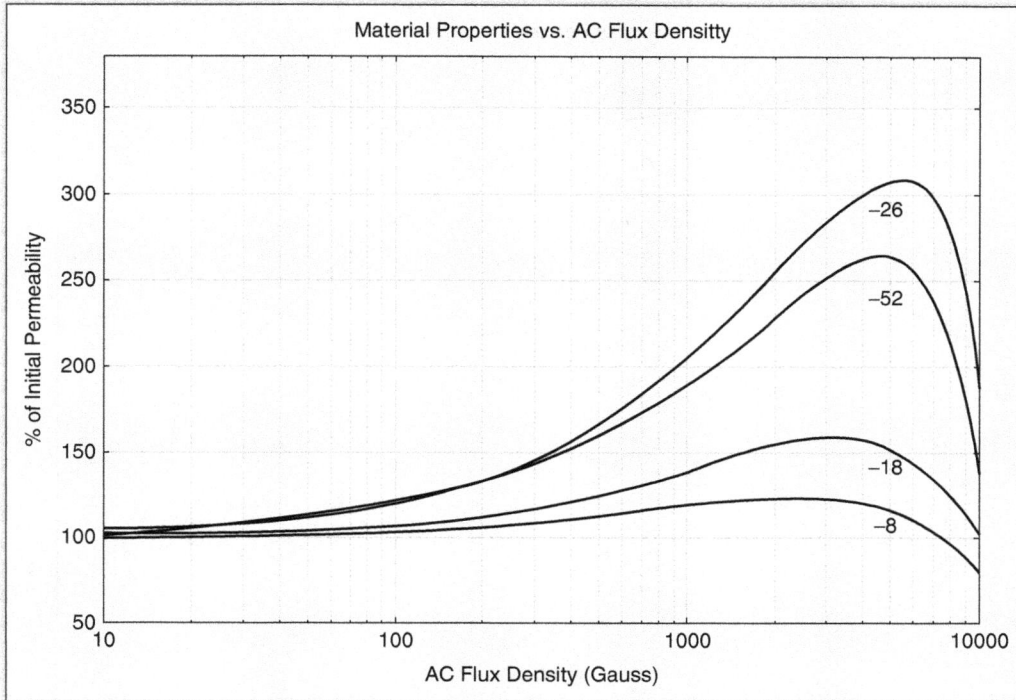

Figure 8-10: Effect of AC flux density on permeability

Micrometals Mix −26 (yellow with white) and −52 (green with blue) are the least expensive of the powder core materials. They also exhibit the largest change in permeability due to environmental factors. Mix −52 is recommended for frequencies above 100 kHz and Mix −26 for frequencies below 100 kHz. If the load for your design requires a large output current range (such as a SSB transmitter supply or an audio power amplifier) and constant inductance is required, then Mixes −18 and −8 are the preferred materials. These have much more stable characteristics with changes in flux, magnetizing force, and frequency. The trade-off is that permeability is lower (55 and 35) and the cost is approximately double for identical-size cores. A larger core is also required, so the final cost of the core can be four times that of a −26 or −52 core.

Table 8-2 indicates the maximum number of turns that will fit in a single layer on various Micrometals toroids. Table 8-3 lists the magnetic path length for

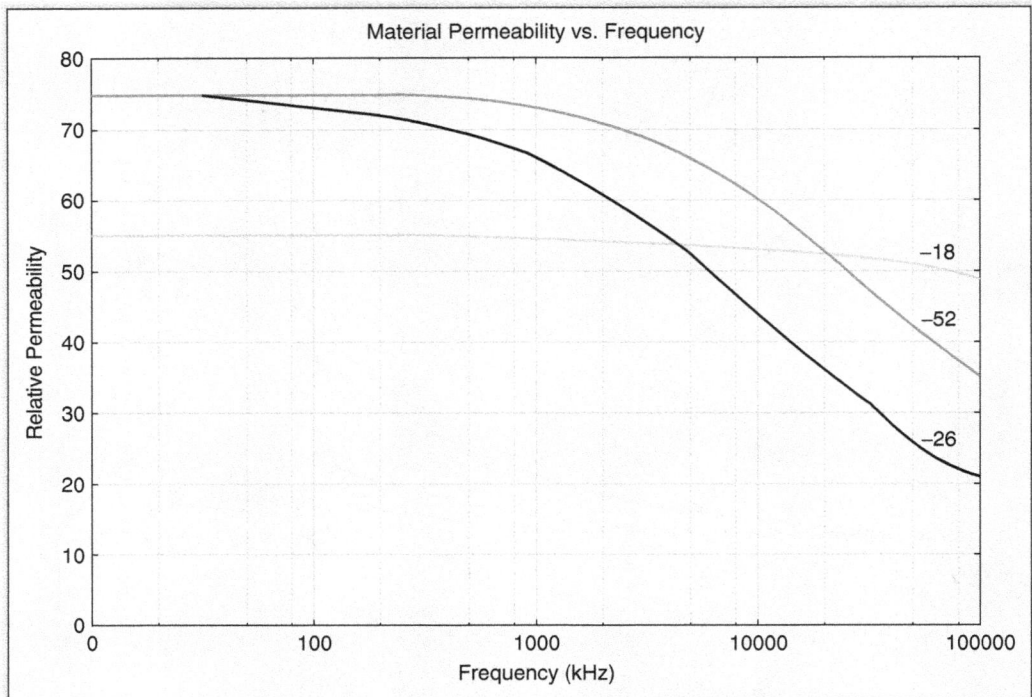

Figure 8-11: Change in permeability due to frequency

various core sizes. We will also have to account for the change in initial permeability due to DC flux. These tables are abbreviated versions of what can be found in Micrometals' application catalog.

Let's do an example of a choke design for a forward converter. The design requires an inductance of 15 μH and has a maximum DC current of 20 A with 2 A maximum ripple. We choose to limit the temperature rise to 40 C° so we can use the current from the 600 A/cm² column of the wire table. This allows an additional 10 C° rise due to core heating. We start with a T-106-26 core to see if a design is possible. We choose this core because it is a convenient size and is the least expensive material. A_L for this core is 900 μH/100 turns. First, we calculate the number of turns required:

$$N = 100 \, (L/A_L)^{1/2} = 100 * (15/900)^{1/2} = 100 * (0.0167)^{1/2} = 13 \text{ turns} \qquad (8\text{-}4)$$

From Table 8-2, we see that 13 turns of #12 wire will fit this core. From Table 8-1 we see that #12 wire will allow 19.6 A. This is close enough to the design

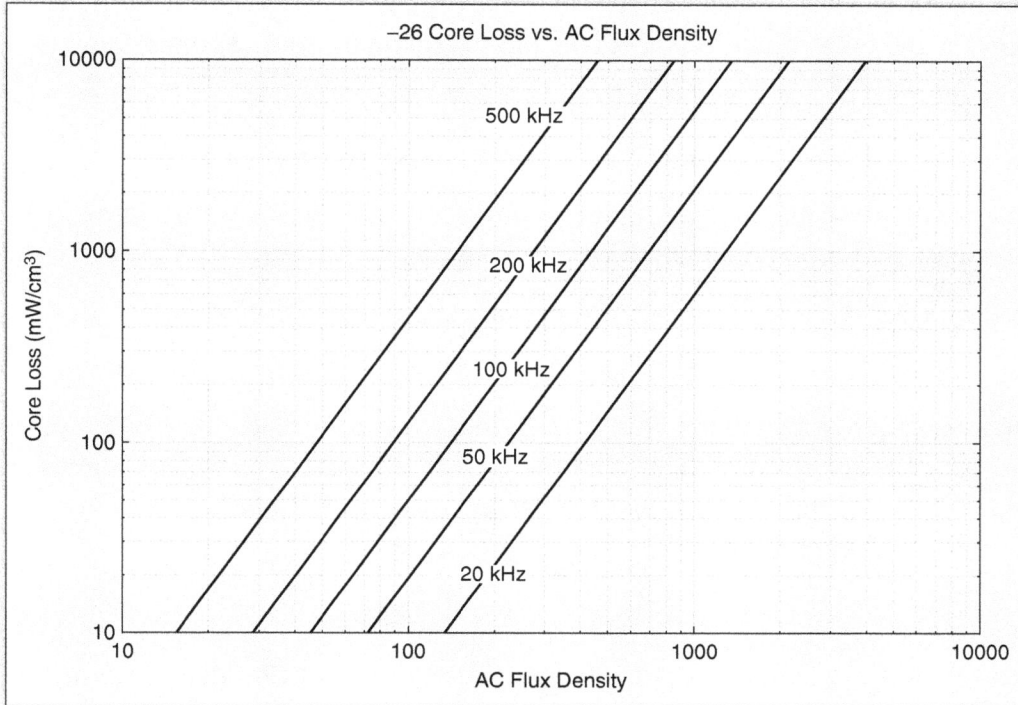

Figure 8-12: Core loss versus AC flux density and frequency for Micrometals −26 material

Table 8-2 Turns for single layer winding versus AWG wire size

Wire Size	T-200	T-130	T-106	T-94
10	33	20	12	12
12	43	25	16	16
14	54	32	21	21
16	69	41	28	28
18	88	53	37	37
20	111	67	47	47

Table 8-3 Magnetic characteristics of various Micrometals toroids

Core	Path Length (cm)	Area (cm^2)	Volume (cm^3)
T-94	5.97	0.362	2.16
T-106	6.49	0.659	4.28
T-130	8.28	0.361	2.99
T-200	13.0	1.27	16.4

goal of 20 A. Now we can verify that the magnetic parameters meet the maximum for the core. Equation 8-5 gives the magnetizing force applied to a core.

$$H = (0.4 * \pi * N * I)/l = (0.4 * \pi * 13 * 20)/6.49 = 50.3 \text{ Oe} \qquad (8\text{-}5)$$

We verify that 50 Oe is below the saturation point, using Figure 8-9.

We can calculate the change in permeability due to DC bias from Figure 8-9. The curve shows 50% reduction of permeability.

Now we can adjust the number of turns based on the reduced permeability:

$$N = 100 \ (L/(A_L * 0.51))^{1/2} = 100 * (15/(900 * 0.50))^{1/2} = 18 \text{ turns} \qquad (8\text{-}6)$$

The next step is to calculate core loss and increase in permeability based on AC flux. Since Figure 8-10 indicates an increase in permeability, we can attempt to use the 13 turns in the belief that the decrease due to DC and increase due to AC will offset each other.

$$B = \frac{2A \times 15 \ \mu H \times 10^8}{2 \times 0.659 \times 13} = 175 \text{ G} \qquad (8\text{-}7)$$

We see that there will be about 30% increase in permeability due to AC flux. If we increase the number of turns to account for the combined effect (50% × 130%), we need 16 turns to get the required inductance. The combination makes 18 turns seem like a reasonable choice. The heating due to the AC flux density can now be calculated, using Figure 8-12.

$$\text{Power} = 120 \text{ mW/cm}^3 * 4.28 = 510 \text{ mW} \qquad (8\text{-}8)$$

The temperature rise can be approximated by:

$$\Delta T = (\text{power in mW/surface area in cm}^2)^{0.833} = (510/26.5)^{0.833} = 11.7 \text{ C}° \qquad (8\text{-}9)$$

Unfortunately, surface area has to be calculated. The manufacturer does not supply this parameter.

We cannot fit 18 turns of #12 wire on this core in a single layer, so we can put an additional two turns of #12 wire over the single layer. This will cause a slightly higher temperature rise but we probably have enough margin because the heating due to AC is close enough to our design goal. It is likely our inductor will rise as much as 45 C° at maximum current.

If we are concerned about the temperature rise, we can redesign the core to lower the flux density and lower the temperature rise in the core. The first step is to choose a mix with lower core loss at the switching frequency. Mixes −52, −8, and −18 all have lower losses at 100 kHz than −26. Since −52 has A_L almost identical to −26, the number of turns will be the same for both materials. For our 18-turn inductor, the power loss is:

$$\text{Power} = 80 \text{ mW/cm}^3 * 4.28 = 340 \text{ mW} \qquad (8\text{-}10)$$

$$\Delta T = (\text{power in mW/surface area in cm}^2)^{0.833} = (340/26.5)^{0.833} = 8.4 \text{ C}° \qquad (8\text{-}11)$$

So we see that changing to a better mix will cut the core loss by 30%. If we wish to cut the loss even more, we can go to the more expensive −18 material. This will require a larger core because this core is already at the maximum turns for #12 wire. Temperature rise is related to both core volume and core area. Notice that core volume increases as the cube of diameter, and surface area increases as the square of diameter. Power loss is directly proportional to volume, so we must reduce AC flux in larger cores to maintain constant temperature rise. Going to Mix −18 will give a larger number of turns to decrease B, but we will have larger core volume. We can see if a T-130-18 core will improve the core loss.

$$N = 100 \, (L/A_L)^{1/2} = 100 * (15/580)^{1/2} = 100 * (0.0167)^{1/2} = 16 \text{ turns} \qquad (8\text{-}12)$$

$$H = (0.4 * \pi * N * I)/l = (0.4 * \pi * 16 * 20)/8.28 = 48.6 \text{ Oe} \qquad (8\text{-}13)$$

74% reduction in permeability due to DC flux.

$$N = 100 \, (L/(A_L * 0.74))^{1/2} = 100 * (15/(580 * 0.74))^{1/2} = 19 \text{ turns} \qquad (8\text{-}14)$$

$$B = \frac{2A \times 15 \text{ μH} \times 10^8}{2 \times 0.698 \times 19} = 113 \text{ G} \qquad (8\text{-}15)$$

$$\text{Power} = 24 \text{ mW/cm}^3 * 5.78 = 138 \text{ mW} \qquad (8\text{-}16)$$

$$\Delta T = (\text{power in mW/surface area in cm}^2)^{0.833} = (138/29.4)^{0.833} = 3.6 \text{ C}° \qquad (8\text{-}17)$$

So we see that we can meet our temperature target with a large margin with a much more expensive core. Another option is to increase the inductance in the Mix −26 inductor so that the ripple current is one-half. This is a potential solution as long as the slower transient response is acceptable for the application.

The numerator of the flux density equation will stay constant (current is one-half, but inductance doubles), so we must increase the number of turns in order to reduce the flux density. The inductance is a function of A_L and N squared. This change will work as long as A_L increases less than 1.414 for the larger core that will be required. The larger core will potentially also increase the loss due to the higher volume to surface area. A new pass at the calculations is the only way to be sure for a T-130-26 core of 30 μH.

$$N = 100\,(L/A_L)^{1/2} = 100 * (30/810)^{1/2} = 100 * (0.037)^{1/2} = 19 \text{ turns} \tag{8-18}$$

$$H = (0.4 * \pi * N * I)/l = (0.4 * \pi * 19 * 20)/8.28 = 57.6 \text{ Oe} \tag{8-19}$$

46% reduction due to DC flux.

$$N = 100\,(L/(A_L * 0.46))^{1/2} = 100 * (30/(810 * 0.46))^{1/2} = 28 \text{ turns} \tag{8-20}$$

$$B = \frac{1A \times 30\,\mu H \times 10^8}{2 \times 0.698 \times 28} = 77 \text{ G} \tag{8-21}$$

$$\text{Power} = 30 \text{ mW/cm}^3 * 5.78 = 173 \text{ mW} \tag{8-22}$$

$$\Delta T = (\text{power in mW/surface area in cm}^2)^{0.833} = (173/29.4)^{0.833} = 4.4 \text{ C}° \tag{8-23}$$

This core works because the A_L value is lower than the T-106-26 core and H is larger for this inductor, so the inductance is decreased even more. In general, though, a larger core will give a larger A_L value, so there is less to gain by going to a larger core. The larger A_L value can keep the number of turns from increasing fast enough to offset the increased volume of the core.

Choosing a Boost Converter Core

The core we just designed for a buck converter is operating in a positive feed-back region. As current increases, the inductance decreases, which will allow more current to flow. The buck converters using the Mix −26 cores will require current limiting to ensure that we do not have destructive feedback. Boost converters generally operate the core at much higher magnetizing force with much smaller inductor values. This makes it more important to evaluate the core for the possibility of saturation under all operating conditions. Designing an inductor in a positive feedback region for temperature will be very dangerous.

Let's choose an inductor for the second boost converter design in Chapter 4. The required inductor is 15 μH at 1.4 MHz with 100 mA ripple current and 450 mA DC current when the input voltage is 2.6 V. The current level is small enough that we can buy an off-the-shelf inductor from a company such as Coilcraft, Toko, or JW Miller. Coilcraft has a very nice calculator on their website that will verify your calculations and point you to a list of appropriate inductors for your design. Coil manufacturers specify the current capability of their inductors in terms of reduced inductance from DC bias and heat rise due to AC flux losses. Coilcraft specifies the saturation current as the point where inductance decreases by 10% due to DC bias and gives heat rise due to AC flux. They give a list of about 50 different inductors that will satisfy our design, including surface mount, axial, shielded, and unshielded. Two shielded surface mount inductors look promising: DS3316P-153 and MSS6132-153. These cores do not show any decrease in inductance due to DC bias.

Higher current boost converter designs will require a custom design. The first boost converter example of Chapter 4 requires 32 μH at 100 kHz with 20.8 A DC bias and 4.2 A AC current. We saw in our design of the buck converter inductor design that it is unlikely that a powder core will handle the AC ripple of this supply without excessive heat. We can choose an E core in Magnetics Type R material for our inductor. This material is a good choice because the permeability increases with rising temperature and with increasing flux density, as shown in Figures 8-6, 8-7, and 8-8. We also see that core loss actually decreases as temperature rises up to 100°C. We can see from the graph of permeability versus flux density that we must keep flux density below 2000 G in order to avoid a decrease in inductance due to saturation.

Magnetics recommends wire with 500 CM/A, which is equivalent to 400A/cm^2 in Table 8-1. Number 10 wire is appropriate for our maximum DC current. We see from the skin effect cutoff equation that #10 wire will show skin effect above 12 kHz, so we will need to factor in additional loss due to skin effect. Magnetics provides a nomograph of A_L versus maximum stored energy for various E cores, as shown in Figure 8-13. The energy stored is proportional to LI^2.

Our design calls for maximum current of 22.9A, so LI^2 is (32 μH * 22.9 * 22.9) or 17 mJ. From Figure 8-13 we see that either a 45021-EC core with A_L

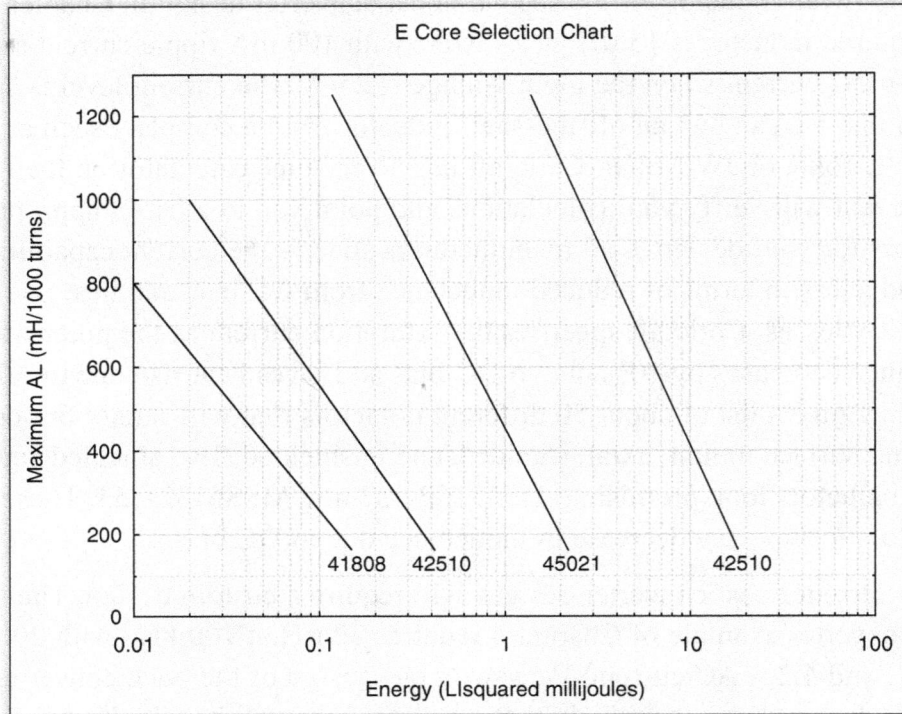

Figure 8-13: Nomograph of A_L versus maximum stored energy for various E cores

below 160 or a 45528-EC core with A_L below 550 will suit our inductor. The data sheet for the 45021 core shows that a 0.1-in. gap will be required for A_L of 160, so this is probably not a good core for our application. The data sheet for the 45528-EC core shows that a gap of 0.025 in. will give A_L of 500. The nomograph of cores versus A_L shows A_L of 400 as a standard gap size. This would be a gap of 0.035 in. for this core and material Type R.

We can now compute the number of turns and other parameters. Note that A_L is mH/1000 turns, whereas it was μH/100 turns for the powder cores.

$$N = 1000 * (0.032 \text{ mH}/400)^{0.5} = 9 \text{ turns} \qquad (8\text{-}24)$$

$$\begin{aligned} B &= (L * \Delta I * 10^8)/(2 * A * N) \\ &= (32 \text{ μH} * 2.1 * 1e^8)/(2 * 3.5 * 9) = 110 \text{ G} \end{aligned} \qquad (8\text{-}25)$$

$$\text{Power} = 0.15 \text{ mW/cm}^3 * 43.1 = 6.5 \text{ mW} \qquad (8\text{-}26)$$

Temperature rise due to AC flux will be negligible.

We can use six conductors of #18 wire in parallel to reduce the skin effect and still have the equivalent current capacity of #10 wire. This inductor will almost certainly have larger parasitic capacitance between the turns. The other option is to twist seven #19 conductors together to produce the equivalent of a single #10 wire. We can take the #10 wire version and the #18 wire versions to the lab and measure losses to see what effect the wire size has on performance.

CHAPTER 9

Transformer Selection

- Transformer Properties
- Safety Concerns
- Practical Construction Considerations
- Choosing a Forward Converter Transformer Core
- Practical Flyback Core Considerations
- Choosing a Flyback Converter "Transformer" Core

Transformer Selection

The basic circuits described in earlier chapters assumed ideal transformers. Real transformers also have inductance, resistance, and capacitance. Losses due to the core material add equivalent resistance. We have the same set of considerations for transformers that we had for inductors. We will have to consider copper and core losses as part of the design to keep the transformer from melting. We will also look at coupled inductors for flyback designs, since they are similar to transformers.

Transformer Properties

An ideal transformer does not allow primary current to flow if the secondary is an open circuit. In actual transformers, there is a small amount of current that flows when the secondary is open. This is the magnetizing current and is a result of the inductance of the primary winding. This magnetizing inductance should be as large as practical so that the magnetizing current is a very small fraction of the current when the transformer is delivering power. When delivering power, the ideal transformer appears as a very small resistance in parallel with the magnetizing inductance. The current through the magnetizing inductance is 90 degrees out of phase with the applied voltage, so no power is consumed by the magnetizing inductance in sine wave operation. Another property of real transformers is that not all of the flux in the core surrounds every turn of both windings. This imbalance creates a leakage inductance in both the primary and the secondary windings.

The choice of core material directly affects magnetizing inductance. An ideal transformer core would have infinite permeability and give infinite magnetizing inductance. Such cores do not exist, of course, so we choose a core with the largest permeability and the least loss for the desired frequency. Figure 8-7 (Chapter 8) is useful for choosing a material for a transformer. Iron powder cores have low permeability, so they are a poor choice for transformer cores. Ferrite cores are the only practical choice for switching supply transformer cores.

The concept of having infinite permeability for a transformer core seems counterintuitive at first. This would cause even a very small AC current to immediately saturate the core. That would be true for an inductor. However, when you remember that the secondary winding creates a magnetic field that is equal and opposite to the primary field, you see that an ideal transformer could transfer infinite power without saturating the core. The limitations in a real transformer are the magnetizing inductance (saturation) and the core loss due to AC flux (temperature rise).

Figure 9-1 shows an equivalent circuit for a real transformer. The leakage inductance of the primary and the secondary and the magnetizing inductance cause problems because we must provide a path for the inductor current when the switch or switches turn off. This is called resetting the core. The capacitance of each winding forms a resonant circuit with the leakage inductances. These resonant circuits cause ringing when the switch opens or closes. A well designed transformer will have almost zero leakage inductances and extremely large magnetizing inductance.

C_P is the primary winding capacitance; R_P is the primary winding resistance; L_P is the primary leakage inductance; R_L is the core loss equivalent resistance; L_M is the magnetizing inductance; C_{PS} is the capacitance between the windings; R_S is the secondary winding resistance; L_S is the secondary leakage inductance; C_S is the secondary winding capacitance.

The capacitance between the primary and secondary allows any AC voltage present from the load (such as high frequency noise from computer circuitry) to be coupled back into the source circuit. A Faraday shield is a simple piece of copper foil that is placed around the primary winding and grounded. The foil completely encircles the primary, but the ends are insulated so that it is not a shorted turn. This creates a capacitor to ground for both windings, but does not

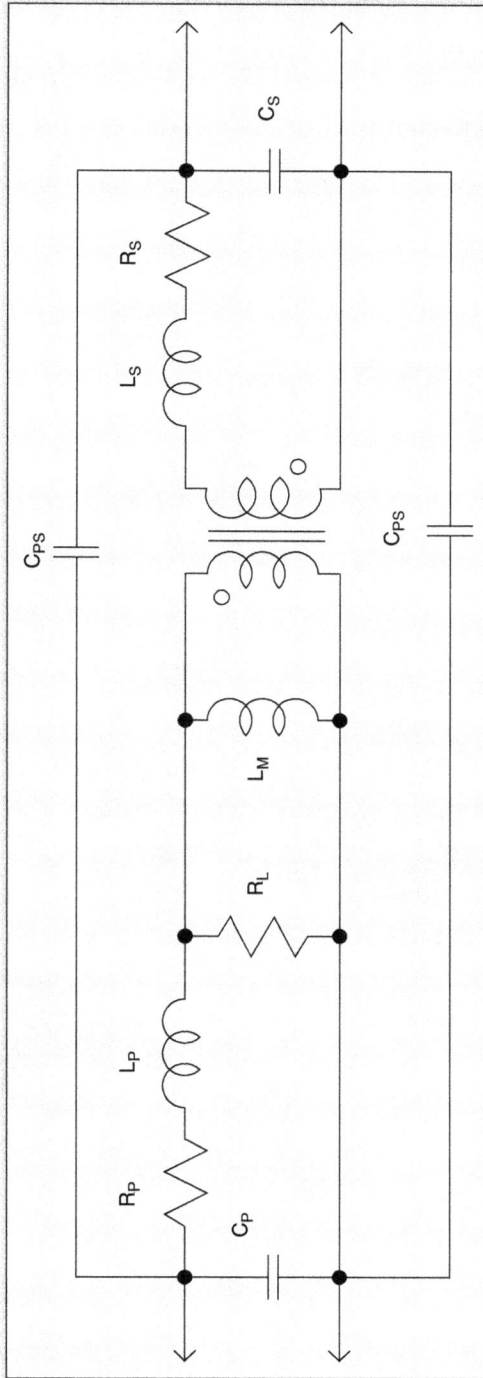

Figure 9-1: Equivalent circuit for a real transformer

affect the magnetic operation of the transformer. The Faraday shield is not always necessary if the EMI filtering is adequate. A Faraday shield loses its effectiveness at higher frequencies because the inductance of the ground lead eventually becomes a significant part of the circuit.

We discussed Faraday's Law in Chapter 5. This law says that voltage is equal to the number of turns times the change in flux:

$$E = N \, d\Phi/dt \tag{9-1}$$

If we rearrange and transform this equation to account for a square wave AC waveform, we get:

$$B = \frac{E}{4 \times f \times N} \tag{9-2}$$

where

E = the peak voltage
f = frequency
N = the number of primary turns
B = the peak AC flux density

We worry about peak AC flux density because it is related to core losses and to saturation of the core. When the AC voltage on the primary increases, the AC flux density increases. The increase in flux occurs because the magnetizing current increases. We must keep the flux from the magnetizing current below the knee of the B-H curve to avoid saturation of the core. The magnetizing current goes down as we increase the magnetizing inductance, which is reflected in Eq. (9-2), with the number of turns being in the denominator—more primary turns increase the magnetizing inductance. Fortunately, frequency is also in the denominator of the equation, so we can reduce the peak AC flux in a transformer by increasing either turns or frequency. Or, conversely, increasing frequency allows fewer turns for a constant flux density. This relationship is what allows us to greatly decrease the size of transformers as switching frequency increases.

Safety Concerns

The primary reason for using a transformer is to provide safety isolation of the load from the power mains. Underwriters Laboratories, Inc. (UL) and the

European safety agencies have specifications for the amount of isolation that is necessary. UL1950 and IEC950 cover safety for information technology equipment. There are other standards that cover applications such as medical equipment or consumer electronic equipment. These specifications primarily concern themselves with the insulation strength between the primary and secondary sides of the transformer and preventing insulation failures from creating a shock hazard. Safety agencies have some strange names for some of the parameters. The ones that we are concerned with in a transformer are creepage (distance along an insulator), clearance (distance through air), and insulation strength (insulation thickness). The insulator around the wire is the basic insulation. Additional insulation is needed because it is possible for a defect in the basic insulation to create a shock hazard. Creepage and clearance specifications exist because air and insulators are not ideal. Contamination can allow small but still dangerous currents to flow through air or along the surface of insulation.

There are two methods of providing enough insulation and fault tolerance shown in Figure 9-2. Figure 9-2(a) shows tape used to provide spacing between the bobbin and the windings so that the combination of spacer and windings provides a solid surface for the reinforced insulation between windings. Reinforced insulating tape is layered on top of this solid layer. The creepage distance is typically 5–6 mm, depending on the safety agency requirements. The spacing tape is one-half of the creepage distance because the current would need to travel along the bottom surface of the insulating tape, around the edge, and then the same distance on the top surface of the insulating tape to the secondary winding. The method shown in Figure 9-2(b) shows the windings composed of triple-insulated wire. The wire is insulated using three different types of insulation so that if any one layer fails, there are still two layers of insulation. If you are not confident in your ability to pass safety agency testing, you should consider using a transformer winding service which is competent in safety agency matters to wind your transformers.

Practical Construction Considerations

Safety agency considerations rule out pot cores and toroids as core shapes because they cannot provide the necessary creepage distances. This leaves all

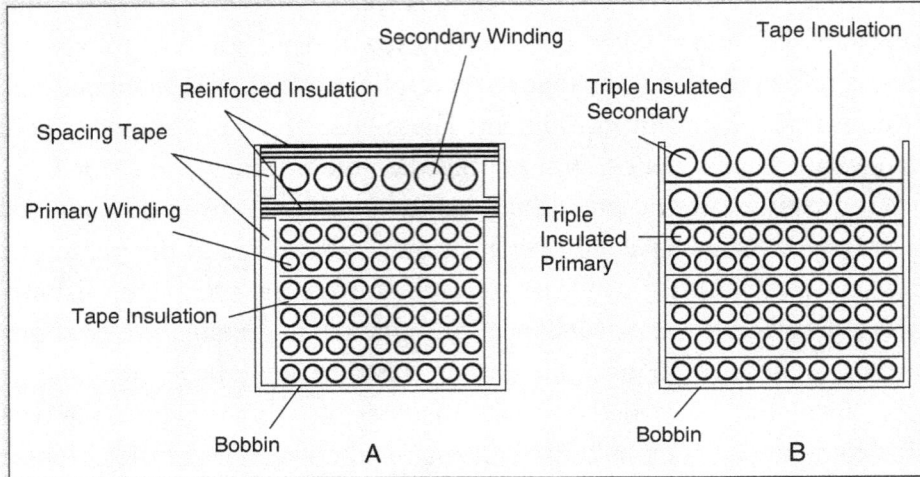

Figure 9-2: (a) Tape used to provide spacing between the bobbin and the windings; (b) windings composed of triple-insulated wire

of the variations of E cores as practical options. An E-type core also allows better heat dissipation for the windings, since a large area of the copper windings is exposed to ambient air. The EC and ETD core shapes have an advantage in high power applications because the round center leg makes each turn 11% shorter than on the square cross section of an E core. This lowers copper loss and reduces the parasitic capacitance. If physical height is a design constraint, then EFD and EPC cores will allow shorter windings than are possible with the other E-type core variations.

Off-line supplies almost always have many more turns for the primary than for the secondary windings. This makes it important to wind the primary closest to the core to limit the total wire length. The parasitic capacitance of a winding is proportional to the surface area of each layer, and inner layers will have smaller surface area. We can also control parasitic capacitance by using Z winding instead of C winding. C winding has a large voltage difference between the ends of the layers. Current due to capacitance will be greater when the voltage is greater. Z winding has each layer wound in the same direction. This causes the voltage differential to be the same between all turns on adjacent layers. Each layer of each winding should be covered with a layer of insulating tape to reduce the capacitance between adjacent layers. Figure 9-3 illustrates Z and C winding methods.

We must consider the wire diameter for transformer use because it can be a source of significant loss. When we looked at choke cores, the wire was usually a single layer for good thermal characteristics. A good rule of thumb for transformers is to use wire with 200–300 A/cm^2 to minimize the heating caused by wire. There is wide variation in recommendations from core manufacturers on an appropriate current density for transformer use. As wire size gets larger and frequency gets higher, skin effect losses rise and can be as significant as bulk resistance losses. Commercial transformers use Litz wire or copper ribbon or copper strip to reduce the effects of frequency. We can use Eq. 8-1 (Chapter 8) to estimate the frequency where a given wire size will have significant skin effect. Above that frequency, the resistance of the wire increases by a factor of 3.2 for every decade of frequency.

MWS Wire Industries makes a line of square cross section magnet wire. The square shape is advantageous for making compact individual layers that stack well. However, the close spacing of the flat surfaces will increase capacitance slightly. MWS also manufactures ribbon and strip. They classify ribbon as having a width up to 0.100 in and strip as having a width of 0.125–2 in. The skin effect is greatly reduced if the conductor is flattened into a thin strip. Many commercial power supplies use strip conductors for the low voltage–very high

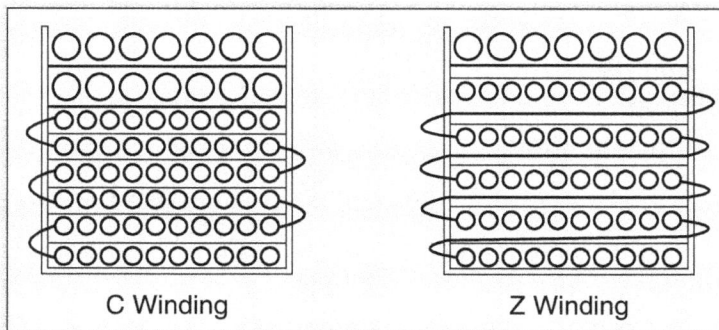

Figure 9-3: C and Z transformer winding methods

current windings such as a 20 A/5 V winding in a PC power supply. These windings are typically only two to five turns.

Polyester tape is typically used to provide safety isolation. 1-mil tape has 5500 V isolation and is rated to 130 C°. 2-mil tape is rated for 7000 V. If higher temperature is required, Kapton polyimide tape is rated to 155 C° and 7500 V. Copper strip is not insulated like magnet wire, so the strip is insulated with tape during the winding process.

Temperature is related to surface area; more surface area gives more heat dissipation (lower temperature rise) for a given loss. Core loss is related to core volume; a larger volume generates more heat for a given flux density. The problem is that volume increases as the cube of the size, and surface area increases as the square of the size. It is interesting to note that the lowest power loss density in ferrites is at temperatures above 25 C°, so elevated temperature operation actually increases efficiency.

The limiting factor for ferrites at frequencies above 20 kHz is temperature rise due to core losses. Below 20 kHz operation, saturation flux density is the limiting factor. A modern supply will operate at least at 20 kHz (more likely, 100 kHz or higher). If you read old articles or application notes (from the era when 20 kHz was a high frequency supply) about choosing a core, you will see references to larger power capacity of cores with bipolar (AC) drive, as in the bridge converter topologies. This is not true for modern high frequency designs because the limiting factor is no longer the saturation flux for the core but, rather, the heat rise in the core due to hysteresis losses. AC drive circuits offer significantly more margin between the maximum flux and the saturation flux than unipolar drives, but even the unipolar drive designs are unlikely to approach saturation flux densities before things start to melt inside the transformer. It is still good design practice to stay below 2000 Gauss to ensure you are well away from saturation.

The last practical consideration is physical size. There is a limit to how small you can make the core based on the amount of wire that must be wound on the bobbin. The amount of current in the secondary windings will dictate the wire size. It may be necessary to choose a larger core just to be able to fit the amount of wire required.

Choosing a Forward Converter Transformer Core

Our example is a transformer designed for a single switch forward converter at 100 kHz with a turns ratio of 5.7:1, input voltage of 310 V and total power of 100 watts (12.6 V at 8.0 A). The ambient temperature maximum is 50 C° and the maximum allowed core temperature is 100 C°. We will actually need two identical primary windings (one for power and one to reset the core) and a secondary winding, as we saw in Chapter 5.

We will choose a Magnetics core. The first step is to look at Table 4 in the Magnetics Design_Application_Notes.pdf available on their website (www.mag-inc.com). This table gives the approximate power handling capability of a large number of cores. E, EC, U, and ETD cores are largely standardized for physical dimensions among manufacturers, so this chart is a useful starting point for cores from many manufacturers. Indeed, other manufacturers, including Ferroxcube, have similar tables. The table lists the 43515 core as suitable for 150 W at 100 kHz. This core is a standard E375-size core that has the same dimensions as standard E-I steel laminations. Choosing this size has the advantage that a different core manufacturer can be used and the same bobbin can be used if a second source for the core material is needed. The next step is to look at the material data sheets for power loss versus temperature. A power transformer will generally have significant temperature rise, so we will need either Type R or Type P material. The temperature specification indicates that Type R material is probably best.

We can start with a three-turn secondary and see if our transformer will work. This indicates that we will have 17 primary turns. We round down the number of primary turns to the next integer. It is better to have slightly more secondary voltage than specified. The next step is to compute the flux density:

$$B = E/(4 * A * N * f * 10^{-8}) = 155/(4 * 0.840 * 17 *$$
$$100 \text{ kHz} * 10^{-8}) = 2713 \text{ G} \qquad (9\text{-}3)$$

Note that E is RMS, so it is P–P/2 for a square waveform.

This is too much flux density for reasonable operation, so we need more turns. Doubling the number will halve the flux density, so we can try 34 primary turns and six secondary turns. 1350 G will generate 100 mW/cm^3 in Type R material. The data sheet shows that the loss will be slightly less as the core temperature rises.

This core has a volume of 5.83 cm^3, so we will have 583 mW of heat from core loss. This will cause acceptable temperature rise. We can get a rough idea of temperature rise by assuming that the perimeter and only one side will be exposed outside of the bobbin. From the data sheet, we calculate that this area is 16.6 cm^2. Using the temperature rise equation,

$$\Delta T = (\text{Power/Area})^{0.833} = (583/16.6)^{0.833} = 19 \text{ C}° \qquad (9\text{-}4)$$

The next step is to choose the wire. The primary will only see about 350 mA, so #28 wire will be adequate at 300 A/cm^2. The current in the reset winding is negligible, but it is easier to wind both the primary and reset windings using the same wire size.

The secondary will require either #14 or #12 wire. Both of these sizes will encounter significant loss due to skin effect. Skin effect begins at 30 kHz for #14 wire and 19 kHz for #12 wire. We can make a rough estimate that skin effect resistance will equal bulk resistance. We will need to use a lower current density to account for temperature rise. Rather than using #12 solid magnet wire, we could twist seven strands of #22 wire to give approximately the same area as #12 but with lower effective resistance. The other alternative is to use a strip conductor that is 0.45 in wide and 0.007 in thick for the secondary. The specifications allow 50 C° rise, so we have 30 C° rise available for the copper loss. We should be well within the temperature rise budget.

The analysis above will be almost identical for a transformer for a bridge circuit. The only difference is that the voltage used in the flux equation will be the full applied voltage because the peak-to-peak voltage is equal to twice the voltage applied in one direction. For a push-pull circuit, we use the supply voltage and the total number of turns of the primary.

Practical Flyback Core Considerations

As with transformers, a ferrite core is the best choice. It is possible to use powdered iron, but there are not many manufacturers that produce cores in E shapes. Another reason is that core loss is generally higher in powdered iron than in ferrite.

All flyback cores will require a gap in order to keep the DC component of flux below saturation. The size of the gap should be kept as small as possible to minimize the fringing field around the gap. A large fringing field contributes to leakage inductance. We wish to store energy in the inductance of the core, but the leakage inductance does not contribute to storage of energy for the output. Another problem with large gaps is that the fringing field will interact with the windings, especially if the secondary is wound with a flat strip. The fringing field will cause eddy currents in the copper and increase losses.

For really low power designs, there are a number of off-the-shelf flyback transformer designs available from Pulse Engineering and Prem Magnetics that are designed to be used with the circuits from application notes from Linear Technology, ST, National, and On Semiconductor.

Choosing a Flyback Converter "Transformer" Core

We will design a transformer for a 60 W flyback converter that will provide 5 V at 5 A, and ±12.0 V at 1.5 A. The design requires an inductance of 4.5 mH with an input voltage of 310 V and operates at 50 kHz with 80 mA of ripple current. The turns ratio is 54.4 for the 5 V winding and 24.0 for the 12 V winding.

We pick a Ferroxcube core. The first step is to choose a core size from their recommended list. This can be found on Page 32 of the Ferroxcube HB2002.pdf that is available on their website (www.ferroxcube.com).

An EC35 core is recommended by both Magnetics and Ferroxcube for the 60 W-level. 3C85 material is a reasonable choice for the chosen frequency. Presumably, we are designing a supply that will go to production, so we can do a preliminary design of the core with ungapped sets and verify its operation in the lab. Then we can order a custom core with a gap for the desired A_L if a standard value does not produce the desired inductance.

Our first constraint is that we must find an integer number of turns for the 5 V and 12 V windings that satisfy the required ratio. This ratio is 54.4/24.0 or

2.267. This is close to 2.25, so starting with four turns for the 5 V winding will give 4 * 2.267 or 9.068 turns. Nine turns will be very close to the desired value. I actually started this example with eight turns for the 5 V winding and 18 turns for the 12 V winding, but when I did the turns calculations, the required A_L was below 100 nH/turn. It was time to do another iteration! We can now calculate the required primary turns.

$$\text{Primary turns} = 4 * 54.4 = 217.6, \tag{9-5}$$

which we round down to 217.

The standard equation for inductance is:

$$L = A_L * N^2, \tag{9-6}$$

so we can rearrange to get A_L:

$$A_L = L/N^2 = 0.0045/47089 = 95.6 \text{ nH/turn.} \tag{9-7}$$

We got really lucky—this A_L value is so close to the standard 100 nH/turn that we will be able to order a standard core set.

We will walk through calculating the required gap to show the method when a standard gap will not work. We start from the A_L for an ungapped core set, which is 2100 for this core. There is an adjustment factor that is applied to the inductance calculation when you are using a gapped core:

$$L * k = A_L * N^2 \text{ or } k = (A_L * N^2)/L \tag{9-8}$$

$$k = (2100 \text{ nH/turn} * 47{,}089)/0.0045 = 22.0 \tag{9-9}$$

The following equation gives k in terms of permeability and magnetic path length:

$$k = 1 + (\mu_i * G/l_e), \tag{9-10}$$

where

μ_i = initial permeability from data sheet
G = gap length in mm
l_e = effective path length of the core

Rearranging gives:

$$G = ((k * l_e) - 1)/\mu_i \qquad (9\text{-}11)$$

so

$$G = ((22.0 * 7.74 \text{ cm}) - 1)/1530 = 0.11 \text{ cm or } 0.044 \text{ in.} \qquad (9\text{-}12)$$

This result is in close agreement with the 0.147 cm gap for A_L of 100 nH/turn from the data sheet.

We can use 0.022-in plastic shims under the center post and each of the outer legs to give the desired gap. We only need one-half of the dimension, since the gap is split into two parts between the center post and each outer leg.

It is important to note that the A_L value given in the Ferroxcube data sheet is nH/turn, Magnetics gives A_L as mH/1000 turns, and Micrometals gives A_L as μH/100 turns. You must be very careful with the dimensions!

Next, we need to use Eq. 9-13 to determine the peak AC flux density in the core.

$$B = (L * \Delta I * 10^8)/(2 * A * N) = (0.0045 * 0.080 * 1e^8)/$$
$$(2 * 0.843 * 217) = 98 \text{ G} \qquad (9\text{-}13)$$

We see from the data sheet that the AC flux core loss for this core will be about 40 kW/m³. Notice that this is equivalent to 40 mW/cm³. Total core loss is 6.53cm³ * 40 mW/cm³ = 260 mW.

$$\Delta T = (260/19.0)^{0.833} = 8.8°C \qquad (9\text{-}14)$$

From Chapter 8, Table 8-1, we can use either #28 or #26 wire for the primary, #16 wire for the 5 V secondary, and #22 wire for the two 12 V secondary windings. Again, it might make sense to use seven strands of #24 wire instead of the #16 wire for the 5 V winding.

CHAPTER 10

A "True Sine Wave" Inverter Design Example

A "True Sine Wave" Inverter Design Example

We will design a "true sine wave" uninterruptible power supply in this chapter. "True sine wave" for our design means 20% or less total harmonic distortion. This product is intended to give instant crossover from line power to battery power and provide power for devices that require sine wave operation. The description in this chapter is intended to show the iterative nature of designing a complicated switching system, so you will see several missteps at each major decision point.

It is imperative that you remember that all of the circuitry in this design is connected directly to the AC power mains. This presents a potentially life-threatening situation. Always use a suitable isolation transformer to isolate the circuit from the AC power mains while testing and analyzing this design.

Design Requirements

The following are the requirements for our design:

1. 115 VAC, 60 Hz, 650 VA maximum input power

2. Class B FCC EMI certification

3. 115 VAC, 60 Hz, 300 VA pseudo-sine wave output, less than 20% total harmonic distortion (THD). Operation over 0.5–1.0 power factor load

4. 300 W-h power capacity

5. instant switchover—zero dropped cycles

Design Description

The requirement for low harmonic distortion means that we will need to approximate a sine with numerous steps. Figure 10-1 shows the spectrum for two waveforms used by square wave converters. The first spectrum is that of a square wave. The amplitude is equal to the RMS value of an equivalent sine wave. The second waveform has four steps that do a reasonable approximation of a sine wave for electronic loads that use full wave rectification. The RMS value is equal to the RMS of an equivalent sine wave and the peak is the same as an equivalent sine wave. Inductive loads, such as motors, need a cleaner sine wave to minimize losses. There are very few capacitive loads in the real world, but an electronic load like a computer will come close. A design that will supply an inductive load will most likely also supply a capacitive load with the same efficiency.

The first design step is to use an arbitrary waveform generator to simulate a stepped sine wave and measure THD. Figure 10-2 shows the test waveforms and the associated spectra for two test waveforms. The spectrum analyzer has a lower limit of 9 kHz, so the test waveforms are 60 kHz. The real design will scale the circuit values by a factor of 1000.

The spectra of the three sampled sine waves shows energy at the fundamental and an alias on each side of the sample frequency. The four-step waveform in Figure 10-1 is sampled at 240 kHz (four samples per cycle), so the first alias frequencies are at 180 kHz and 300 kHz. These are the same frequencies that occur for a simple square wave. The first waveform in Figure 10-2 is sampled at 480 kHz (eight samples per cycle), so the alias frequencies occur at 420 kHz and 540 kHz. The second signal in Figure 10-2 is sampled at 960 kHz (16 samples per cycle), so the alias frequencies are at 900 kHz and 1020 kHz. There are no harmonics of the 60 kHz fundamental anywhere in the spectrum for the signals of Figure 10-2. All the energy at higher frequencies is related to the sample frequency. Energy decreases for each alias as frequency increases. Table 10-1 lists the energy relative to the fundamental for each alias up to 3 MHz for the eight-step signal. The total harmonic distortion (THD) is 4.6%, which is well within the 20% THD goal. This means we can drive the load directly without filtering. The 16-step waveform has about 1% THD.

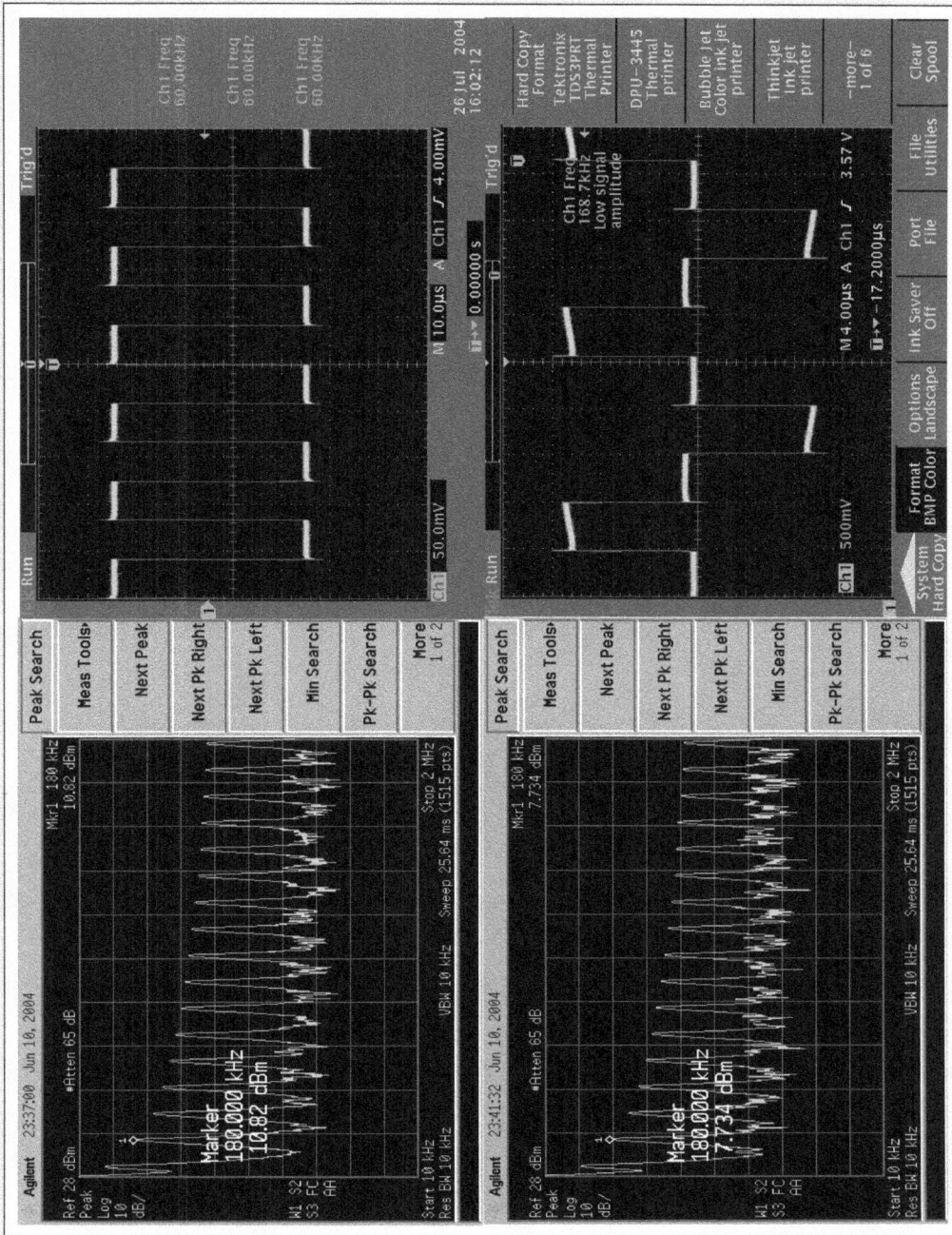

Figure 10-1: Spectra for two waveforms used by square wave converters

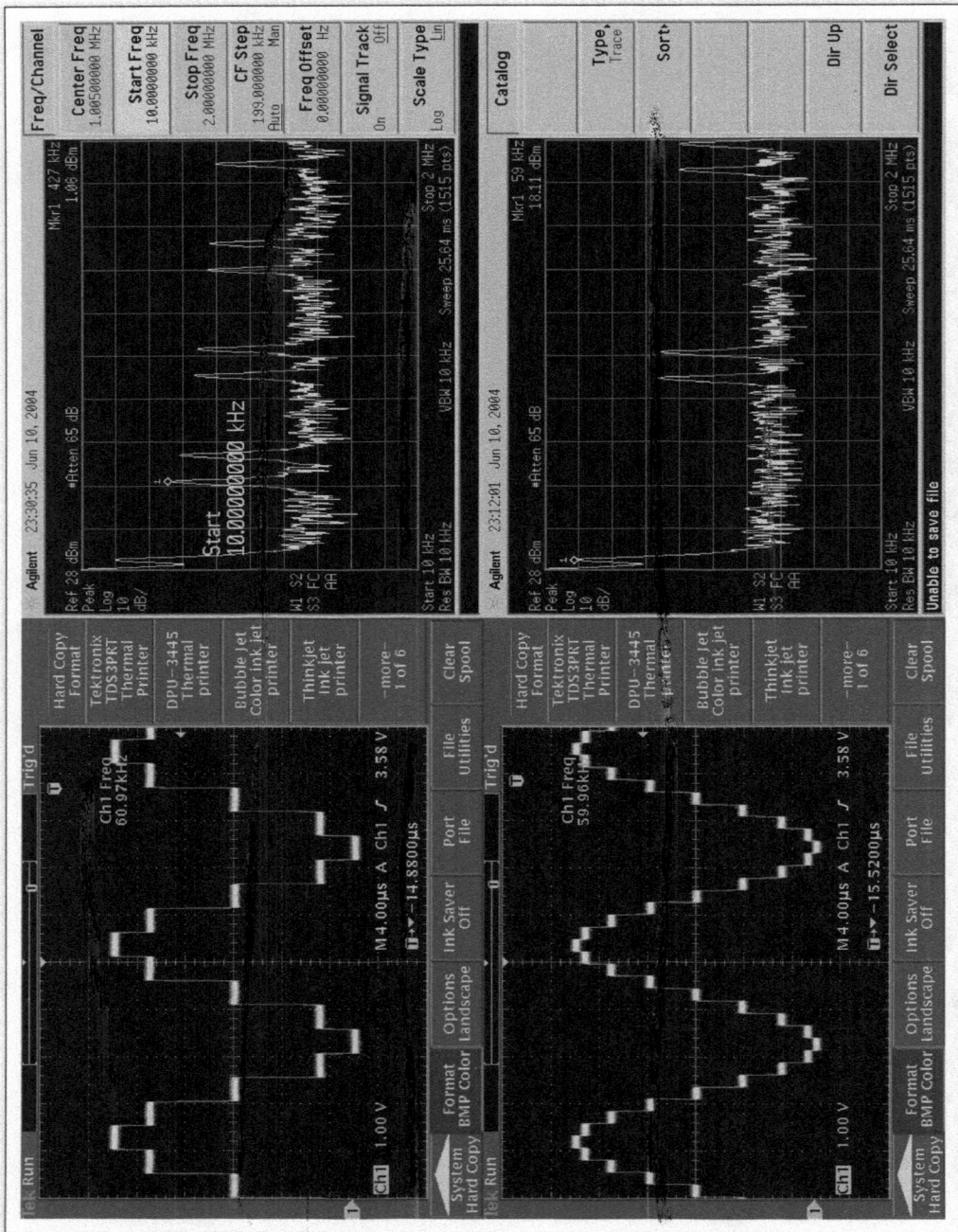

Figure 10-2: Test waveforms and their associated spectra

Table 10-1 Relative Distortion Power versus
Alias Frequency for Eight-Step Waveform

Frequency (kHz)	Relative power
420	0.020
540	0.013
900	0.004
1020	0.003
1380	0.002
1500	0.002
1860	0.001
1980	0.001
2340	0.000
2460	0.000

The next decision we must make is how to switch from mains power to battery power. We could use a relay to provide switching from mains power to battery power. Relays are not especially fast devices, so we would need to provide a reservoir capacitor to hold enough energy for the period that the relay is switching. It is not long, but tens of milliseconds are normal. We also require electronics to control the relay. If we use electronics to provide the switch, we will likely have approximately the same number and cost of components as the relay solution. Another reason to look at an electronic solution is that the relay has reliability issues. Current flow causes contact wear on the normally closed contacts and the normally open contacts oxidize over time. An advantage of the electronic solution is that we can likely produce a circuit that will power the output from both the input supply and the batteries while the input supply reservoir capacitor is being depleted. An even simpler and significantly less expensive transfer circuit uses diodes to isolate the battery pack from the main power supply. The output power conversion circuit will pull up to 15 A from the battery pack. A Schottky diode will have about 0.5 V forward voltage, so the power dissipation will be approximately 8 W. An electronic solution will likely have power dissipation near 2 W. The analysis of the battery pack reveals that there is no advantage to the extra complexity of the electronic transfer because there is ample reserve energy in the battery pack.

Sealed lead acid batteries are the only cost effective technology in this application. This market is very competitive, so prices are almost identical among manufacturers for equivalent batteries. We will almost certainly want at least

24 V, and perhaps as much as 48 V, to minimize the current draw. We must derate the capacity of a battery based on peak current draw. A 20 Ah battery is capable of supplying 1 A for 20 hours. The same battery will supply 20 A for only about 36 minutes. Our 300 W-h requirement means that the 20-hour rating must be 500 W-h. Table 10-2 lists several configurations and total cost (Digi-Key prices in case lots):

We have four candidate battery packs that will supply the required energy with roughly equivalent cost. With a difference of only $3.30 between the highest and lowest, we need to look at complexity and reliability to choose an appropriate system. The 36 V system appears to be the best choice because it gives a significant margin in energy. It also will cost less to manufacture because it uses three identical batteries, where the least expensive option mixes 12 V and 6 V batteries.

A lead acid battery has a float voltage of 2.40–2.42 V per cell when fully charged. Our system will require 44 V across the battery pack to maintain charge. Allowing a 6 V drop across the charging circuit means we need 50 V for the charging circuit and the power conversion circuit. Likewise, a lead acid battery is significantly discharged when the voltage drops below 1.95 V per cell. This sets our lowest power conversion voltage at 35.1 V.

My original design goal was to use as many off-the-shelf components as possible. The current levels in the preregulator preclude using a single off-the-shelf inductor in the output filter. I ran a second design using the 72 V battery. This quick analysis revealed that the duty cycle would range from 50% to almost 100%. The current level decreases, but the change in duty cycle requires

Table 10-2 Battery Composition versus Total Cost (2004 Dollars)

Battery voltage	Total voltage	Capacity (Ah)	W-h	Total cost
6	120	4.2	504	141.75
6	72	7.2	518	135.64
6	42	12	504	115.45
12	72	7.2	518	97.34
12	48	12	576	106.82
12	36	17	612	98.64
12	24	28	672	99.90
12/6	42	12	504	96.60

600 µH at 7 A rather than 470 µH at 12 A. The current level for both inductors will require a special-order part.

The nominal output voltage is 120 VAC. The highest supply voltage needed corresponds to the 170 V peak of the AC output voltage. The second supply voltage is 120 V. We can use a boost converter or a transformer converter to step up the voltage from the battery. As mentioned in Chapter 4, the boost converter cannot limit current under fault conditions because the switch is not in the current path between input and output. A forward or push-pull converter allows us to shut off the output by shutting off control of the switch.

There are two options for supplying the two voltages. The first is to use a single output and use the PWM circuit to change the voltage between 120 V and 170 V. The second method is to use the converter to generate both voltages simultaneously and electronically switch between the voltages. For the first method, the PWM control loop must be fast enough to track the change from 120 V to 170 V during the 2-ms time period that the voltage needs to have the peak value. It is possible to design a control loop that will respond so quickly, but it is likely to be quite complex.

The second method uses more components but the control loop is much simpler. Our design will generate both the 120 V and 170 V sources from separate windings on the power transformer. We can only control one of the voltages, so we must provide a way to ensure the two voltages always have close to a 50 V difference. One solution is to control the 170 V supply and clamp the 120 V to 50 V below the 170 V supply. This is probably a reasonable solution, since the 120 V supply will provide the bulk of energy in a true sine wave application and will tend to drop below 120 V. In an electronic application, the 170 V supply will provide the majority of the energy.

The final stage to consider is the output driver. The output section is a standard H bridge that generates the alternating AC signal. The biggest difference between the H bridge here and an H bridge in a switching power supply is that the voltage and current may very well not be in phase. Out-of-phase operation is certain for a load with less than unity power factor. This requirement for current out of phase with voltage requires that we use MOSFETs so that current can flow in either direction in the on switches.

Preregulator Detailed Design

Figure 10-3 shows the schematic of the input power conversion circuit. Only one-third of the mechanical power switch is shown. The first section controls power to the power line supply. The second section disconnects the battery from the output circuit and the third section disconnects the low voltage tap of the battery. The input inductors and the differential mode capacitors are designed to increase the power factor to the greatest extent possible within a reasonable cost. Economical inductors will not be effective at reducing low-order harmonics. There are no harmonic requirements at this time in the United States, so reducing harmonics is not a requirement. We can reduce the harmonic content by reducing the size of the input reservoir capacitor and increasing the ripple voltage.

Figure 10-4 shows the input regulator that reduces the input voltage to 50 V. The peak battery voltage during float is 43.6 V. 50 V gives enough head room for the battery charging circuit. Keeping the input voltage low reduces the range for the final power conversion circuit. The lower difference between the input voltage and the float voltage reduces losses while charging the battery pack. The battery charger is a dissipating regulator, so higher input voltage will waste more power. Setting the maximum duty cycle for the input regulator to 90% allows the lowest voltage to be 56 V, and the 187 V maximum input will set the minimum duty cycle to 26%. We can calculate the reservoir capacitor using the 108 VAC minimum input.

We use the hold-up equation from Chapter 3 (Eq. 3-1) to determine the minimum input reservoir capacitor. The hold-up energy can be approximated by

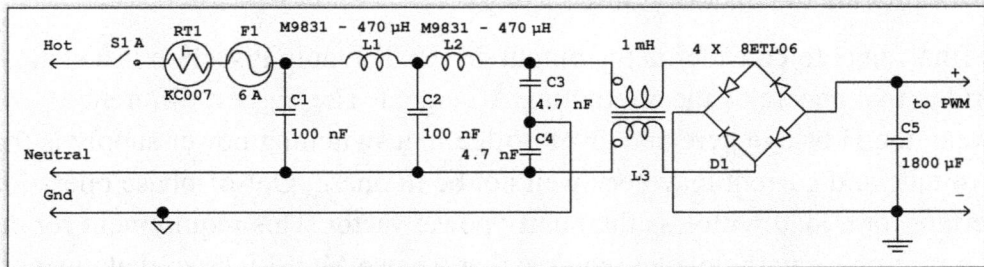

Figure 10-3: Input power conversion circuit

286

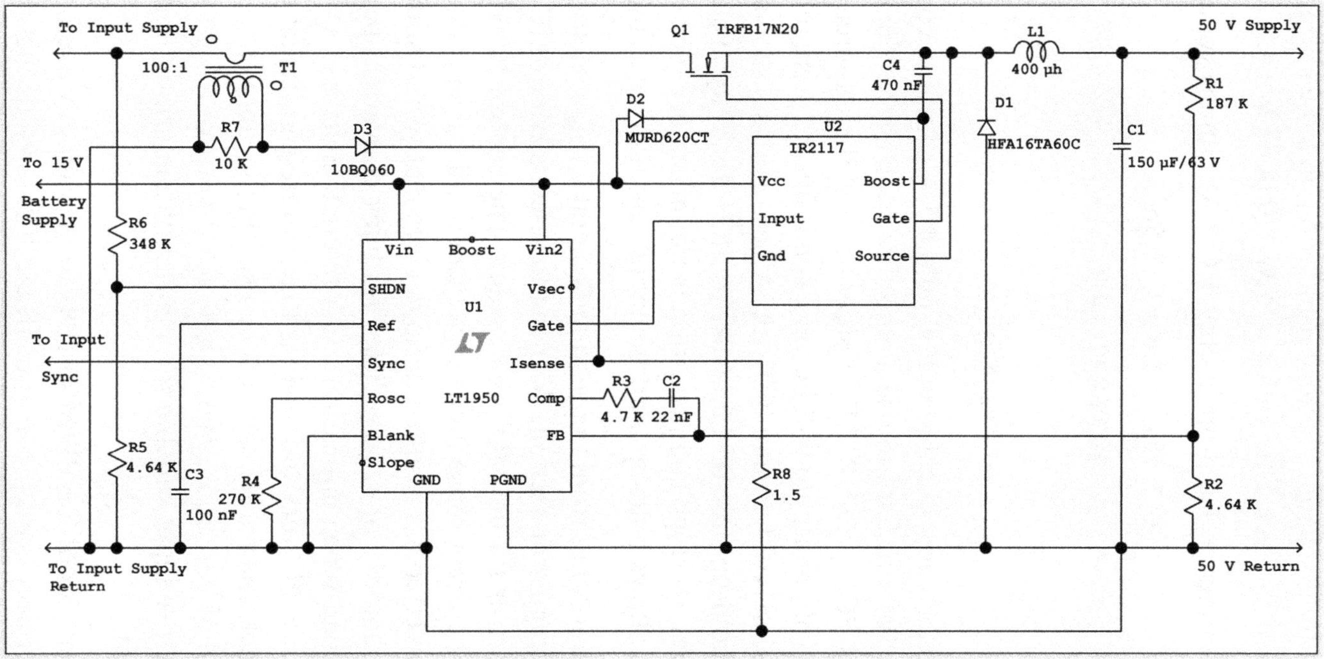

Figure 10-4: Input regulator that reduces the input voltage to 50 V

using the entire time for one-half cycle (8.3 ms). We need to supply both charging current for a fully discharged battery pack (4 A) plus the current to provide the full output power (7.5 A). This yields 575 W * 0.0083 s = 4.8 J.

$$\text{C * Peak Voltage}^2 = \text{Hold-up Energy} + \text{C * Minimum Voltage}^2 \quad (10\text{-}1)$$

$$\text{C} * 152^2 = 4.8 + \text{C} * 56^2 \quad (10\text{-}2)$$

$$19{,}968\ \text{C} = 4.8, \quad (10\text{-}3)$$

so

$$\text{C} = 240\ \mu\text{F} \quad (10\text{-}4)$$

A search for a 240 μF capacitor that will handle 4.5 A of ripple current reveals no available parts. We will have to use a capacitor such as the Panasonic 1800 μF 200 WV model ECOS2DP182EX in order to meet the required ripple current capacity. The voltage ripple will be very small for the preregulator, but harmonic suppression will be more difficult. A new analysis shows 13 V ripple at low input with hold-up time of 7.5 cycles to 56 V. The duty cycle range only needs to be 26–50%, so slope compensation is not necessary. Setting maximum duty cycle to 50% allows hold-up until the input capacitor discharges to 100 V.

Next, we use the algorithm from the section General Design Method in Chapter 4 to design the preregulator section. We need to select a control IC that can drive a high side driver for the MOSFET switch. The control IC will also need to have an external current sense input so we can use a current transformer. A search of buck converter ICs resulted in no devices that would allow use at 200 V input. The vast majority of devices listed for buck converter operation were aimed at point of load-type applications with internal switches. A search of off-line ICs found mostly very old designs that require significant design work. A forward or flyback control IC can also be used in this application. The LTC1950 meets all of our requirements: gate drive to drive a high side drive circuit; a current sense input suitable for a current transformer; and internal slope compensation.

We will want to sync all of the PWM circuits to the output waveform to control EMI. This does not necessarily reduce EMI, but it assures us that the levels will be constant over time. The control IC has a range of 100–500 kHz. We will want to use a crystal to control the frequency, so we need a frequency that is conveniently divided from a standard crystal frequency. 122,880 Hz is 256 times 480 Hz and 12.288 MHz is a standard frequency. 491,520 Hz is 1024 times 480 Hz and 4.9152 MHz is a standard frequency. Both of these frequencies will require a divide by 10 circuit and multiple binary divisions. 4.9152 MHz yields 153 kHz for switching frequency, so it is slightly better than the 122 kHz for the 12.288 MHz oscillator. 153 kHz is a 6.54 µs cycle time.

The duty cycle range is only 2:1, so transformer drive is feasible, but a high side driver IC is likely to be less expensive and more straightforward. The IR2117 is a good choice. The IR2117 will only drive 250 mA peak current, so we will have to verify that switching time is adequate. The required current capability of the high side drive depends on the gate charge of the MOSFET. The IRFB17N20 is reasonable because it has low on resistance and is the least expensive of the low r_{ds} 200 V MOSFETs. It has 30 nC of gate charge. The switching time will be 120 ns (30 nC/250 mA). A 470 nF capacitor will supply all of the current necessary for the boost circuit.

The current sense transformer can be a special design or we can use a standard toroid with a single turn primary. The secondary will not carry appreciable current, so a miniature toroid (around 1 mH) should be adequate. R7 needs to be large enough to generate sufficient volt-seconds to reset the core. R8 will be determined empirically in the lab to generate the required voltage at 7 A peak inductor current. A reasonable start from the 100:1 guess is 1.5 Ω. The feedback resistors need to divide the 50.0 V output to 1.23 V. The actual values are arbitrary. We chose 4.64 K for R2, so R1 must be 187 K. The resistor divider on the shutdown pin sets the shutdown voltage to 100 V.

The next step is to pick the ripple current and design the inductor. This design is a cascaded system, so we need to use a longer transient response for one section than the other. The load on the input section will tend to vary less than the output load, so we can use a small amount of inductor ripple. The low ripple current will provide a long response time. Ripple current of 500 mA is a reasonable value. We use the inductor equation to calculate the inductor value:

$$L = V * dt/di = (187 \text{ V} - 50 \text{ V}) * (0.26 * 6.54 \text{ µs})/0.50 \text{ A} = 466 \text{ µH} \quad (10\text{-}5)$$

Our prototype will use four 390 µH (Miller parts from Digi-Key) in series/parallel to result in 400 µH at 11 A and 780 µH at lower current. The ripple current at full output will be around 600 mA instead of the desired 500 mA, and the ripple current will be 300 mA at light loads. A 150 µF/63 V Panasonic EEUFC1J511 can handle 690 mA ripple current and has an impedance of 0.178 Ω. The ripple voltage will be on the order of 70 mV.

The HFA16TA60C is an adequate commutation diode. The average current at full load will be 8.5 A and we need more than 200 V PRV to have enough margin. This diode blocks 600 V and has 16 A average current capability. D2 will have almost the full 187 V input when reverse biased, so an MURD620CT 200 V FRED will be appropriate. The second diode of the package is tied in parallel. D3 will have a short high voltage pulse during core reset, so a 10BQ060 60 V Schottky diode will be adequate.

We do not need an auxiliary power supply to run the control IC. The battery pack provides an "instant-on" power supply for the control IC. A single battery will not provide enough voltage to drive the switch when the battery is fully discharged, and two batteries supply too much voltage when the batteries are fully charged. We will use two batteries and reduce the voltage to provide the IC control voltage. We do not need soft start operation, since the load will be supplied by the battery until the output reaches the battery voltage.

Output Converter Detailed Design

Push-pull operation is a reasonable choice for the output converter, since it will derive its input from a relatively constant input voltage. The frequency doubling allows using much smaller inductors and filter capacitors. The National LM5030 is an excellent choice for this application. It meets all of our requirements: external synchronization, internal slope compensation, and large gate drive capability. We will set the maximum duty cycle (input side) to 40% to leave room for control and to stay away from the need to ensure that the devices do not conduct simultaneously. The lowest voltage is 35.1 V battery voltage minus the 0.7 V voltage drop of the switching diode, or 34.4 V. The maximum voltage is the 50 V preregulator voltage minus the 0.7 V drop of the switching diode, or 49.3 V. This sets the minimum duty cycle to 28%. The duty cycle on the output side will be double or a range of 56–80%.

The RMS AC output current is 2.5 A. However, the average current for the 170 V supply will be 900 mA and the 120 V supply will be 1.25 A. The peak currents will be 3.5 A and 2.5 A, respectively. The maximum rectifier diode voltage will occur at minimum duty cycle. The maximum input voltage for a 170 V supply will be 303 V plus the rectifier voltage drop. The maximum 120 V supply will be 214 V. PRV for both supplies will equal double the input voltage. One way to reduce the stress on the diodes for the 170 V supply is to put a 50 V supply in series with the 120 V supply instead of a single 170 V supply. This arrangement will force the two supplies to track more closely. Increasing the maximum duty cycle to 45% and putting two supplies in series gives 190 V and 79 V for the two PRV values. The minimum duty cycle changes to 63%. The HFA08TA60 is a reasonable diode for the 120 V supply, but the HFA16TA60 is actually less expensive and allows us to use the same part in two positions. The MURD620CT is a reasonable diode for the 50 V supply. The forward voltage drop at 1 A forward current is 1.2 V for both diode types, so the transformer voltages need to be 80 V and 191 V. Figure 10-5 shows the circuit for the power conversion circuit.

The output current is low even at full load, so we can start with 600 mA ripple current. The supply will transition to discontinuous mode at 300 mA output current, which corresponds to about 25 VA in the load. The 50 V inductor needs to be:

$$L = V * dt/di = (79 \text{ V} - 50 \text{ V}) * (0.63 * 3.27 \text{ μs})/0.60 \text{A} = 100 \text{ μH} \quad (10\text{-}6)$$

The 120 V inductor needs to be:

$$L = V * dt/di = (190 \text{ V} - 120 \text{ V}) * (0.63 * 3.27 \text{ μs})/0.60 \text{A} = 240 \text{ μH} \quad (10\text{-}7)$$

Putting the supplies in series makes the peak current the same in both supplies, so both inductors need to be rated for 3.8 A peak current.

The lowest input voltage is 34.4 V. The switch current is on the order of 12 A, so the voltage drop for the IRFB33N15D is on the order of 0.7 V. The current sense resistor will drop 0.5 V. The lowest transformer voltage is then 33.2 V.

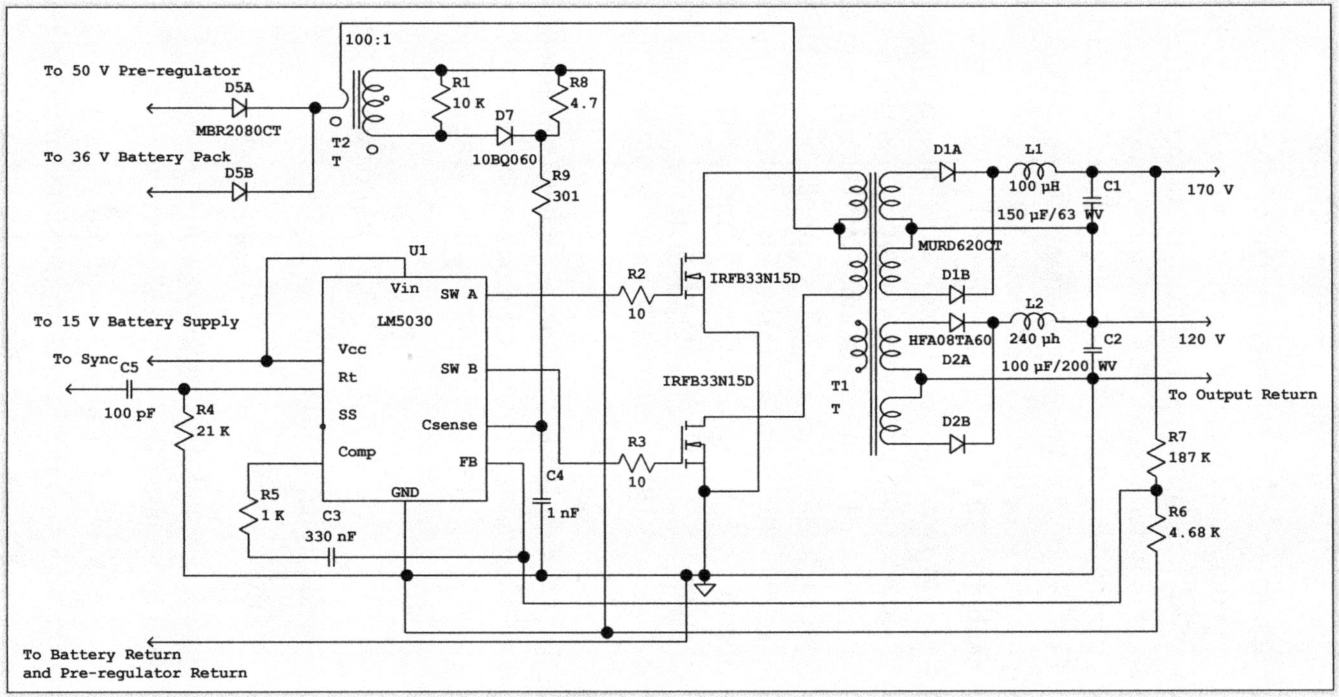

Figure 10-5: Circuit for the power conversion circuit

The transformer ratios for the windings are 80/33.2 = 2.41 for the 50 V supply and 191/33.2 = 5.75.

A reasonable set of windings will likely be four turns per side on the primary (eight turns total) of #10 wire. The 50 V secondary could be nine turns per side of #18 wire. The 120 V winding could be 23 turns per side of #18 wire. This is a reasonable place to start with our prototype.

Once again, the first design decision needs correction. Since the LM5030 has a 500 mV current sense voltage, the resistor would dissipate almost 6 W. This indicates that a current sense transformer is a more reasonable design.

The IRFB33N15D is rated at 33 A drain current at 25°C. It can still handle the 12 A peak current at 150°C. The drain voltage rating must be 150 V because the maximum drain voltage will be double the 50 V input voltage. A 100 V device will not have any margin.

The output ripple voltage is not especially important in this application. The Panasonic 100 µF/200 WV EEU-EB2D101 has enough ripple current capability and a low dissipation factor. The 150 µF/63 WV EEU-FC1J151 is reasonable for the 50 V supply. The MBR2080CT is a good power transfer diode for D5. The average current at full load will be 11 A and we only need 60 V PRV to have enough margin. Only one diode in the package will conduct at any time.

H Bridge Detailed Design

The output H bridge is a standard design, as shown in Figure 10-6. We will have enough margin with 200 V MOSFET switches. The high side switches can be driven with an IR2117 such as we used in the input preregulator.

The 1 µF boost capacitors for the IR2117 drivers should be film capacitors so they have minimal leakage. Electrolytic capacitors are not acceptable because of the amount of time that the high side drive must be on. The high side drive for the 170 V supply can use a 60 V P channel MOSFET because it must only withstand the voltage of the 50 V supply. C3 supplies speed up current to turn on Q5 because the current supplied by R1 is insufficient by itself to turn on the

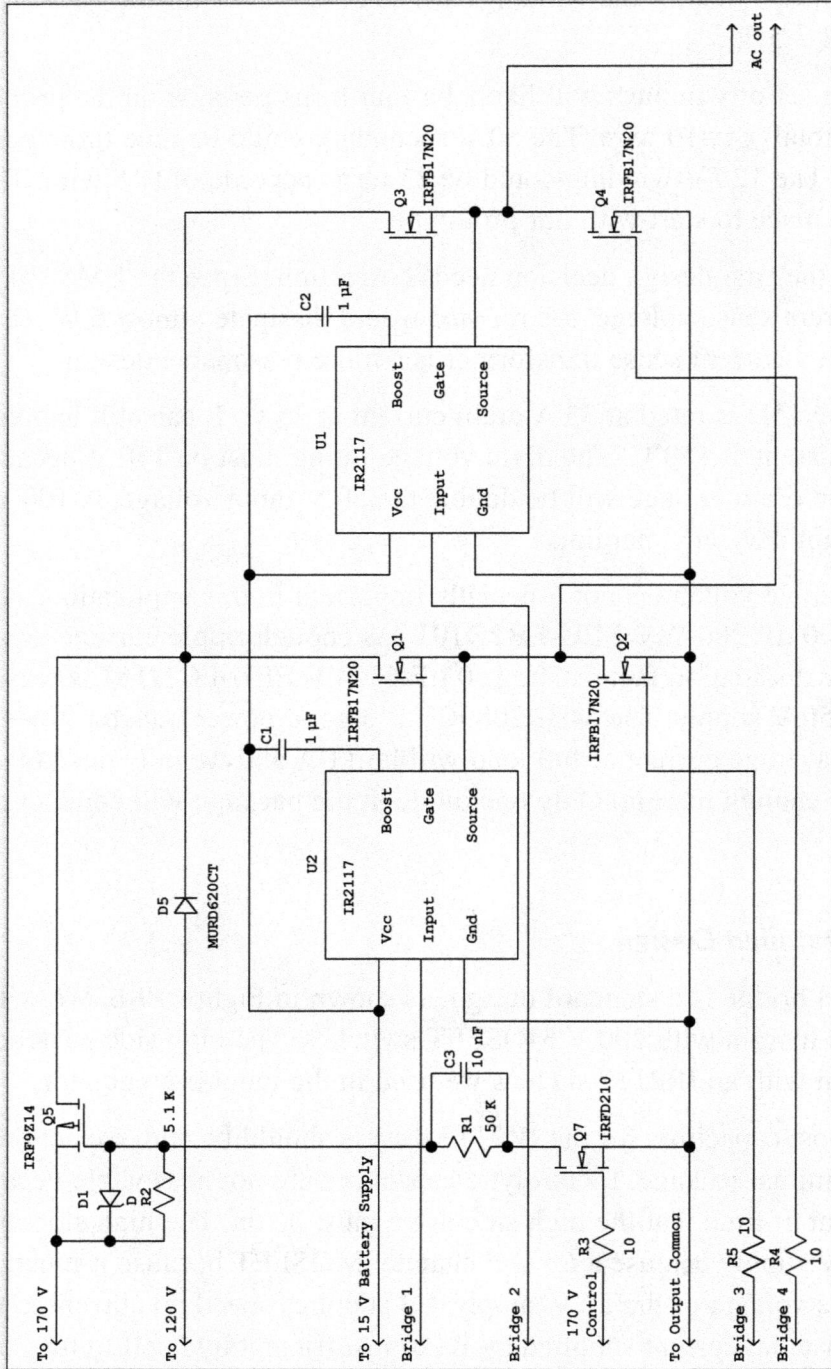

Figure 10-6: Output H bridge

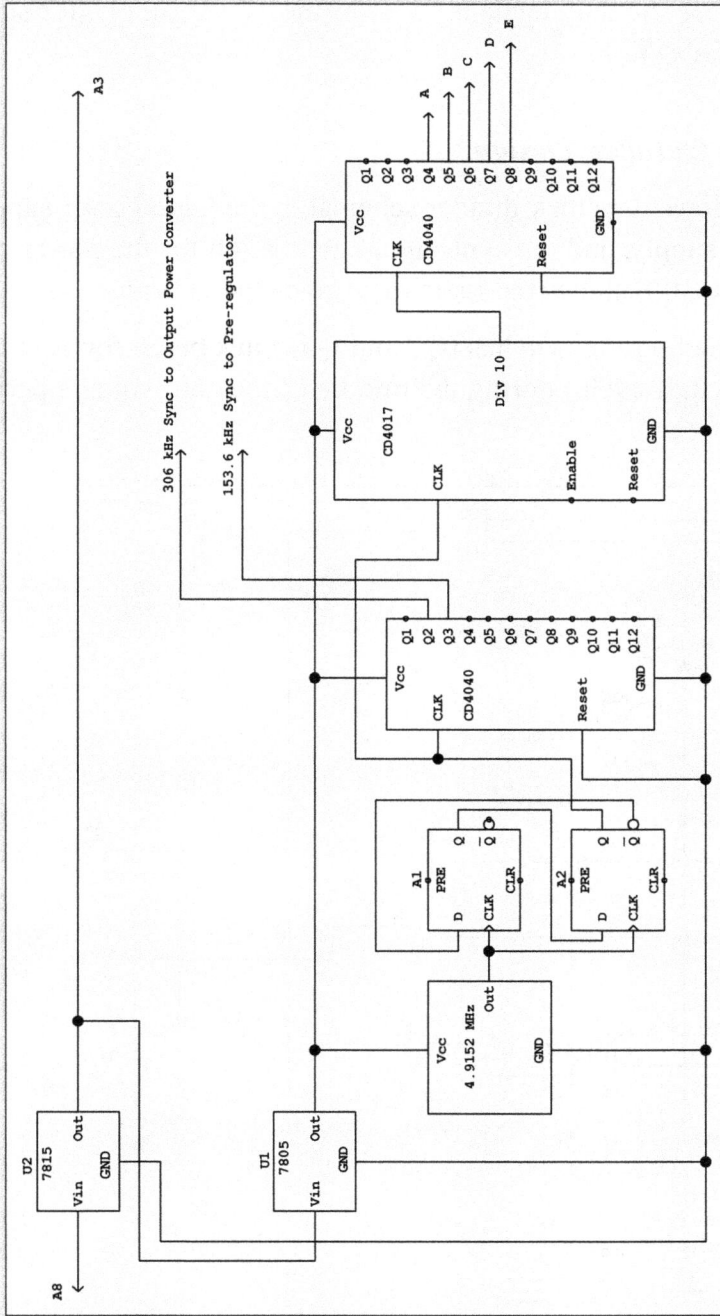

Figure 10-7: Schematic for the clock and battery regulator

switch. R1 must be a 500 mW resistor. The resistor will only dissipate about 250 mW because the duty cycle is 25% for the 170 V supply. R2 supplies the turn-off path for Q5.

Bridge Drive Detailed Design

Figure 10-7 shows the clock divider schematic, the logic power supply, the battery regulator supply, and the sync signal generation for the power conversion circuits. Figure 10-8 shows the logic that drives the H bridge.

The drive to the bottom switches (Q2 and Q4) must be on for 1.04 ms (one-eighth of each half-cycle) during the middle of the zero voltage period. This

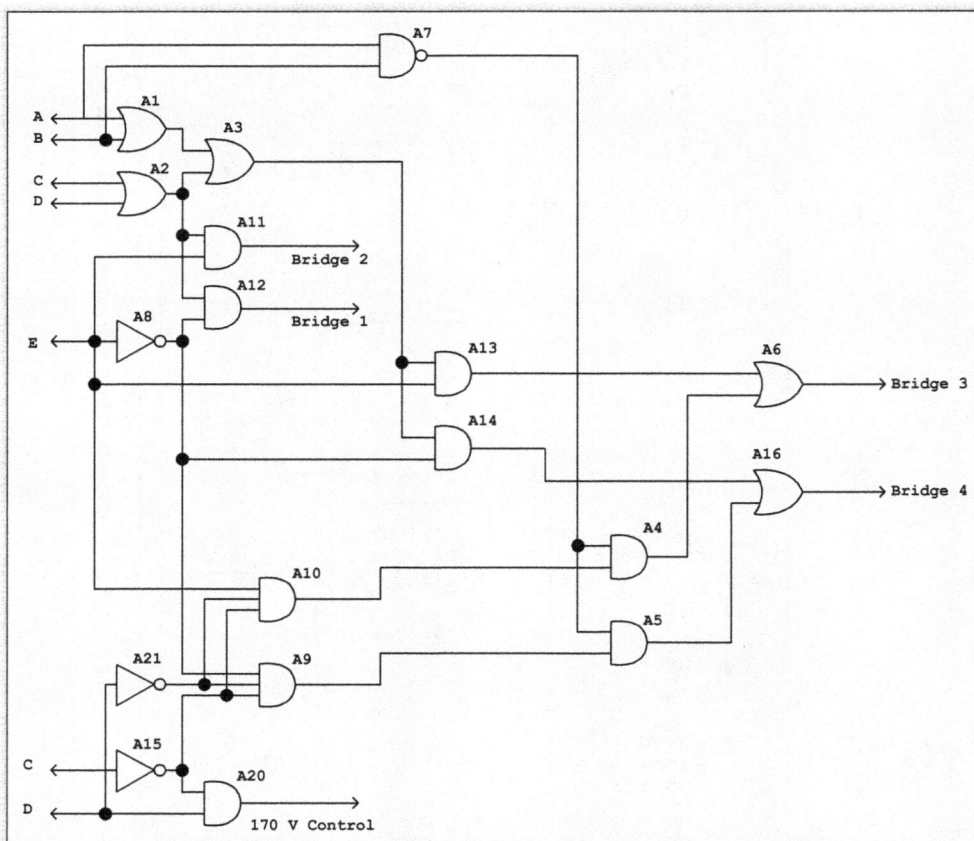

Figure 10-8: Drive logic for the H bridge

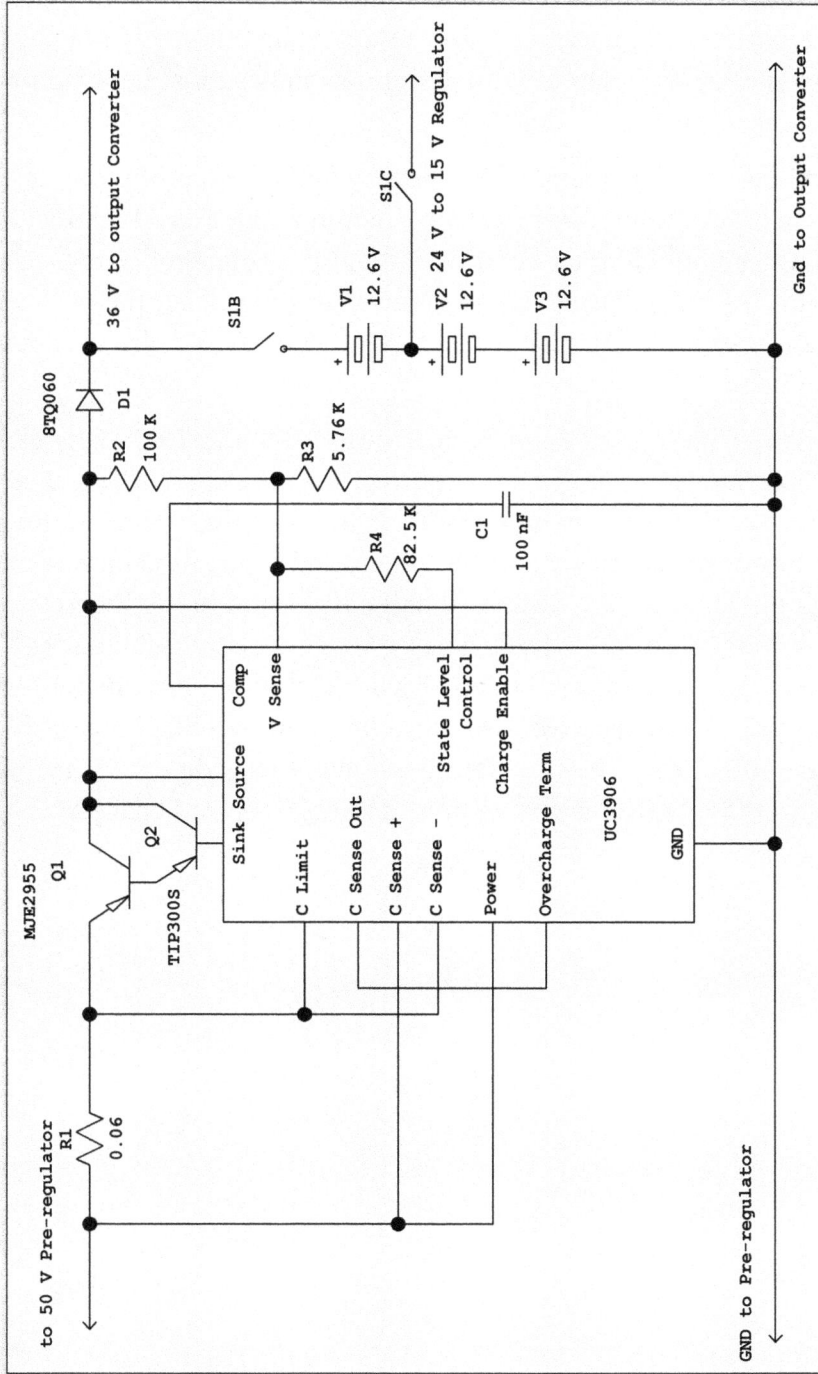

Figure 10-9: The battery and its charging circuit

297

sequence allows the current to be steered from one bottom switch to the other during the zero time and guarantees no overlap of the bottom switches and the top switches. The top switches and the bottom switches must have nonoverlapping drive to ensure no current shoot-through.

Figure 10-9 shows the schematic of the battery and the charging circuit. The charging circuit follows the design method recommended in TI Application Note U-104. Q1 and Q2 implement a darlington pass element because the IC can only supply 25 mA of drive. D1 is required to disconnect the IC from the battery when power fails. D1 guarantees that none of the IC circuit is reverse biased by the battery.

You should notice that the schematics in Figures 10-4 and 10-5 show a long connection from the GND pin of the control IC to the output filter capacitor. This is a reminder that the current sense resistor and components on the left side of each of the schematics should be connected to the GND pin of the IC and the GND pin should have a single small connection to the PGND connections of the power side of the circuit. The power line circuit and the 50 V preregulator should be built on one PC board with the output circuits built on a second PC board. The alternative is to build the input circuits on one side of the PC board and build the output circuits on the other side with very wide power traces in the middle for connection to the battery and charger.

CHAPTER 11

A PC Off-Line Supply

- Setting Requirements
- The Input Supply
- DC–DC Converter
- Diode Selection
- Inductor Designs
- Capacitor Designs
- Transformer Design

A PC Off-Line Supply

In this chapter we will do a paper design of a 220 W PC switching power supply and examine all the steps necessary to create a working design. It is important to note that the schematics presented here are the result of a design analysis and are not a working design. The intent is to show the steps necessary to start from paper and go into the lab to finish the design.

It is imperative that you remember that almost all of the circuitry in this design is normally connected directly to the AC power mains. This presents a potentially life-threatening situation. Always use a suitable isolation transformer to isolate the circuit from the AC power mains while testing and analyzing this design.

Setting Requirements

In general, the output requirements are set by the needs of the system and are fairly inflexible. For a PC, this is typically.

5.0 V(+/−5%)	18 A	50 mV ripple
12 V(+/−5%)	14 A	120 mV ripple
−12 V(+/−10%)	0.5 A	120 mV ripple
3.3 V (+/−5%)	14 A	50 mV ripple

Search for ATX12V_1_3dg.pdf on the web for an example of a complete power supply specification. Table 11-1 gives a checklist of the options that must be selected and our example decisions.

This list represents our initial cut at specifications. It is entirely possible at this stage that these cannot all be satisfied at the same time.

Table 11-1 Power Supply Options and Decisions

Parameter	Selection
Voltage selection (manual/automatic)	Automatic
Power line frequency (50/60/universal)	Universal
Power factor correction (yes/no)	No
Power voltage range (10%/15%/20%)	20%
Voltage hold-up time	2 cycles
Line regulation	0.5%
Load regulation	0.1%
Current limiting (constant/foldback)	Constant
Maximum internal temperature	60°C
Expected ambient temperature	10–50°C
Mean time to failure	>100,000 hours (11 years)
Overall efficiency target (maximum load)	80%
FCC Part 15 class	B
Total component cost (per 1000)	$50.00

The Input Supply

Figure 11-1 shows a typical universal input power supply. It has a power switch, a fuse, an RFI filter, an NTC inrush current limit thermistor, a full wave bridge rectifier, and a filter capacitor.

The worst case for our design is a brown-out on a 115 VAC 50 Hz system followed by a 2-cycle power loss. The peak line voltage during a −22% brown-out is:

$$115 * 0.78 * 1.414 = 127 \text{ V} \tag{11-1}$$

Our requirements call for the power supply to continue to provide the entire 220 W of output even when the power line is dead for 40 ms. From Chapter 3, we see that the worst-case time is approximately 48 ms from the peak of the input waveform.

We need two pieces of information to choose the capacitance of the input supply. The first is the energy that will be drawn during the 2-cycle power loss. The second is the range of input voltage for the DC–DC converter. This value is rather arbitrary at this point. Making this voltage low (e.g., 80 V) will allow us to use a smaller reservoir capacitor on the input for potentially lower cost. Such a low voltage complicates the task of designing the DC–DC converter

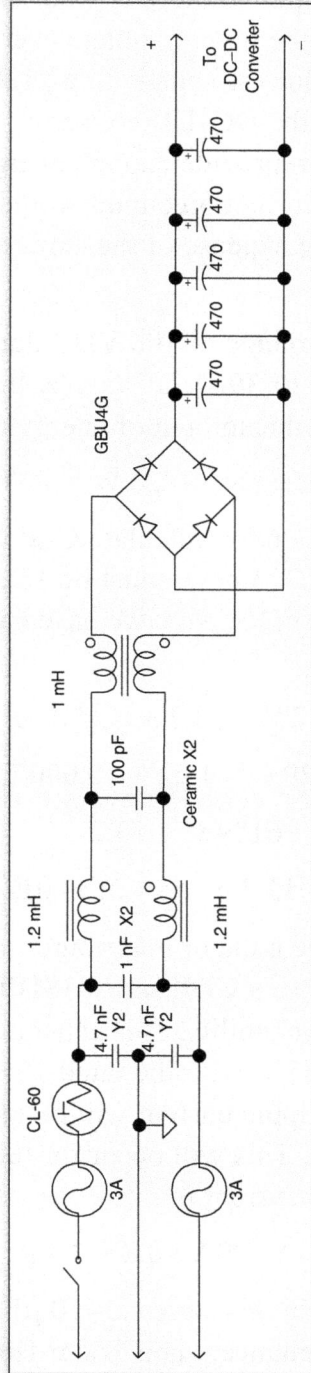

Figure 11-1: A typical universal input power supply

because the ratio of potential input voltages is 4.85. The larger input range will make it harder for it to control the output voltage over the entire range of power line conditions. Increasing the lowest voltage (e.g., 120 V) will reduce the complexity and improve control of the DC–DC converter, but it will require as many as five or six capacitors to provide enough energy for the power loss condition. An active power factor correction circuit would make the DC–DC converter design much easier at the expense of the extra components for the preregulator.

We will choose the minimum voltage of 100 VDC during a power outage. Our requirements yield a maximum of 390 VDC (set by the maximum expected line voltage). First, we need to find the amount of energy used during power loss:

$$\text{Energy} = (P * T)/\text{efficiency} = (220 \text{ W} * .048 \text{ s})/0.80 = 13.2 \text{ J} \qquad (11\text{-}2)$$

We can use simultaneous equations to calculate C, since the energy storage at the peak of the 90 VAC (127 VDC) input must be 13.2 J more than when the voltage on the capacitor is 100 VDC. Now we have enough information to calculate C:

$$C * 127^2 = 13.2 + (C * 100^2) \qquad (11\text{-}3)$$

$$16,129 \text{ C} = 13.2 + 10,000 \text{ C} \qquad (11\text{-}4)$$

$$6129 \text{ C} = 13.2 \qquad (11\text{-}5)$$

$$C = 13.2/6129 = 2154 \text{ } \mu\text{F} \qquad (11\text{-}6)$$

The input filter capacitor can be a major contributor to failure if it is subjected to large ripple current. We choose a CDE series 381EL capacitor to satisfy our margins for voltage range and reliability. This capacitor will give more than 100,000 hours of operation at 65°C with the rated 105°C ripple current. As a first approximation, the RMS ripple current will be identical to the RMS current drawn from the power line. This will occur for the 90 VAC brown-out condition, so our worst-case ripple current is:

$$(220 \text{ W}/90 \text{ V})/0.8 = 3.1 \text{ A} \qquad (11\text{-}7)$$

The largest capacitor available in this series is 470 µF. Paralleling five of these will satisfy our minimum capacitance. (This is our first clue that we may not be able to meet both the cost target and the hold-up time target). This capacitor is rated for 1.4 A ripple current at 120 Hz, so we have 7.0 A of ripple current capacity. We have to derate this value by 10% for 100 Hz operation, which yields 6.3 A capacity. We have plenty of margin for variation beyond the design ambient temperature and design current. This value also gives us margin for each capacitor to account for unequal currents. The capacitor with the largest capacitance will also have the largest ripple current. As mentioned in Chapter 6, the majority of loss of capacitance occurs below −20°C, so we do not have to account for reduction due to temperature.

The bridge rectifier can be a garden variety bridge, such as the General Semiconductor GBU4G. The data sheet indicates that this bridge can run at 3.1 A RMS and 240 VAC without a heat sink if we provide 0.47 in. × 0.47 in. of copper foil for the leads on the PC board or if we supply a small aluminum heat sink. The Thermometrics CL-60 in-rush current limiter is adequate to protect the GBU4G. It has a cold temperature of 10 Ω, so the maximum in-rush current will be about 27 A for a 240 V system.

We choose nominal values for the X and Y Line filter capacitors. The 4.7 nF Y capacitors will give worst-case 0.5 mA leakage in a 240 V 60 Hz system. The 0.1 μF and 100 pF X capacitors are a good starting point. The values of the common mode and differential mode inductors are also reasonable starting points. The values of these six components will need to be adjusted based on the results of EMC compliance testing.

DC–DC Converter

Figure 11-2 shows the DC–DC converter schematic. My preference for all supplies is a current mode controller. The pulse-by-pulse current control guarantees that there will not be any destructive current through the switch. The UC1842 series is produced by multiple manufacturers, including Linear Technology and Texas Instruments (formerly Unitrode). Both manufacturers have ICs with improved performance compared to the original series. These ICs are intended for off-line and high power DC–DC converter applications. We will use the Linear Technology LT1241 because it has duty cycle internally

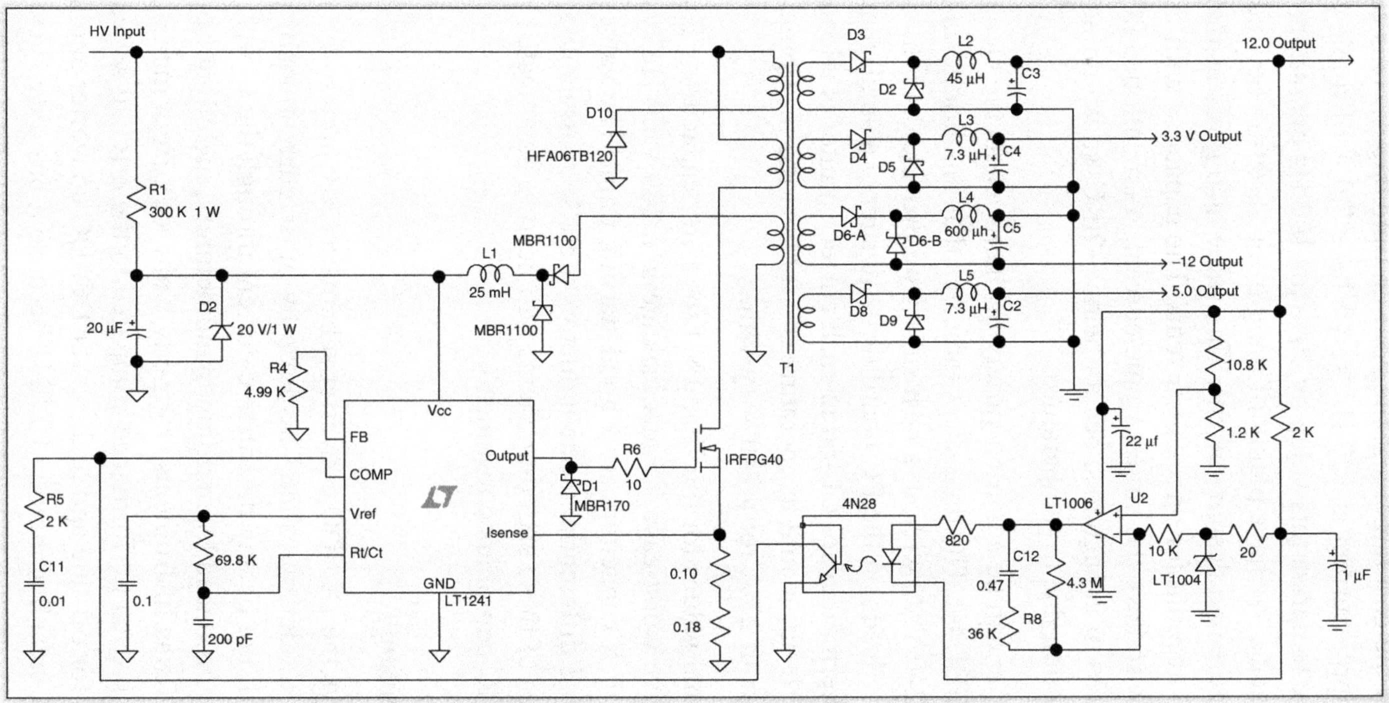

Figure 11-2: DC–DC converter schematic

limited to 50%. Much of the background information for this design comes from the LT1241 data sheet, Linear Technology Application Note 25, and Unitrode Application Note U-100A.

A single switch forward converter will work for our design. The flux in the transformer will need to be reset each cycle, so we will need a controller that has a 50% duty cycle limit. This topology constrains our choice of switch to a 1000 V device. We will choose an IRFPG40 HEXFET. This device has an I_{DMAX} of 4.3 A at 25°C, so this will fit with our worst-case RMS current of 3.1 A. We will verify that this device is really adequate later in the design process.

The next thing we need to select is the operating frequency. We will choose 100 kHz. The input voltage has a very large range, so we will want as little dead time as possible. We will select the timing capacitor and resistor from the nomographs on Page 6 of the LT1241 data sheet. We select 200 pF to give dead time as close to 0% as possible. Now we find that the resistor must be 70 kΩ for 100 kHz operation.

The next step is to design the bootstrap power supply for the LT1241. We start by choosing the trickle supply resistor (R1). The LT1241 draws approximately 250 µA while in the under-voltage lockout state. We must supply this current, plus more, to charge up the power supply capacitor (C1). The startup time is controlled by how fast R1 can raise the voltage on the capacitor to 9.6 V. On the other hand, charging the capacitor more quickly will require a higher power resistor for R1, which raises the internal temperature of the power supply and reduces efficiency. We will choose to allow 0.5 W of dissipation at the highest input voltage.

$$R = E^2/P = (390^2)/0.5 = 305 \text{ K} \tag{11-8}$$

The closest standard 5% value is 300 K.

We choose to run the IC at 10 V to supply adequate gate charge and to give enough margin above the under-voltage cutout. During load transients, the auxiliary supply may temporarily increase above 10 V. In order to protect the IC, we provide a zener clamp to limit the supply to 20 V. The controller IC only draws 8 mA at 10 V for proper operation. The bulk of the current consumption comes from the gate drive for the MOSFET switch. We can calculate the required gate drive current from the FET data sheet. We will drive the FET with 10 V at 100 kHz. Figure 6 of the data sheet shows that we will need approximately 90 nC of gate charge for each cycle. We have 100,000 cycles per second, so the charge is:

$$90 \times 10^{-9} * 10^5 = 0.009 \text{ C/s, or } 9 \text{ mA} \tag{11-9}$$

We choose the value of the current sense using data from the LT1241 data sheet and the maximum drain current for the FET. The maximum drain current that the FET will handle is limited by the case temperature. We will allow the FET to rise 25°C above the ambient or 85°C. This limits the drain current to 3.4 A and sets the size of heat sink required for the FET. The following formula from the LT1241 data sheet gives the value of the current sense resistor:

$$R_s = 1.0V/I_{Peak} = 1.0/3.4 = 0.294 \ \Omega \tag{11-10}$$

The control of the main 12 V output must cross the isolation barrier. The circuit containing U2 and the 4N28 optoisolator provides the feedback necessary to control the output and the isolation needed for safety. This circuit is an exact duplication of the feedback circuit used in the off-line example in the Linear Technology Application Note 25. The LT1006 compares the output voltage against the 1.2 V reference provided by the LT1004 and drives the LED of the 4N28 in proportion to the difference. The 4N28 LED is returned to the voltage reference to keep the op-amp referenced sufficiently above ground. The compensation network composed of R5 and C11 is based on a rough estimate (taken from the Application Notes) of the values necessary for proper operation. The 4.99 K resistor from the LT1241 feedback pin to ground ensures that the internal error amplifier does not draw down the control voltage at the compensation pin. The 4N28 transistor regulates the voltage at the compensation pin to maintain output voltage control. The values for R5 and C11 will be adjusted based on performance testing of the prototype. They will almost certainly need to be adjusted for good transient performance. R8 and C12 also provide compensation of the feedback loop and may need to be adjusted in the lab.

The LT1241 output driver has a peak current output of 1.0 A. Resistor R6 provides current limit protection for the IC and is equal to the power supply voltage divided by the peak current limit of the IC. D1 prevents transient voltages coupled by the FET internal capacitances from taking the output pin more than a Schottky diode drop below ground. Large negative transients can cause instabilities in the IC because the parasitic diode in the output transistor and substrate will be turned on.

Diode Selection

We need to know the maximum voltage on each winding in order to know what peak reverse voltage is required. The ratio of lowest to highest voltage is 390/100, or 3.9. All of the diodes I have selected are produced by International Rectifier.

The main 5 V output will have 10 V reverse voltage at low input and 3.9 * 10.8 = 42.1 V at high input. This voltage is within the capabilities of Schottky diodes. Standard reverse voltages for Schottky diodes are 30 V, 45 V, and 60 V. 45 V is too close to the peak voltage to allow for transients; we need to use a 60 V diode. The MBR4060WT is a dual diode in a TO-247 package that will suit our needs. D8 and D9 will each use both diodes in the package. The data sheet shows that the expected forward voltage is 0.7 V. The device will dissipate no more than 0.7 * 18 or 13 W. An Aavid model 532802b02500 heat sink will handle 13 W with a 50°C temperature rise and no forced airflow.

The 12 V output will have 24 V reverse voltage at low input and 3.9 * 25.2 = 98.3 V at high input. This voltage is just barely within the capability of a 100 V Schottky diode, but it leaves no margin for handling transients. A better choice is to use an ultra-fast diode, even though it will dissipate more power. A suitable device is the MUR2020CT dual diode in a TO-220 package. D2 and D3 will each use both devices in a package. It can handle the 14 A average forward current at 150°C case temperature. The data sheet indicates that this device has an expected forward voltage of 0.9 V at 14 A. This device will dissipate 0.9 * 14 = 13 W. An Aavid model 532802b02500 heat sink will handle 13 W with a 50°C temperature rise with no forced airflow.

The −12 V output has the same parameters as the +12 V output except that the expected current is significantly lower. We will still need an ultra-fast diode to have enough margin at the highest input plus transients. A suitable device for D6 is the MURD620CT dual diode in a D-Pak package. It can handle 0.5 A with a case temperature of more than 150°C. This device is surface mounted, but if a large area of copper is supplied, there will be adequate heat dissipation. The forward voltage at 0.5 A is 0.8 V. The device will dissipate 0.8 * 0.5 = 0.4 W. The data sheet indicates 80°C/W from ambient to the junction. We can expect the junction temperature to be 80 * 0.4 + 60 = 92 C°, which is well within the capabilities of the device.

The 3.3 V supply is similar to the 5 V supply, but the peak voltage will be 3.9 * 7.1 = 27.7 V. A Schottky diode will easily fit for our needs. A suitable device for D4 and D5 is the 45 V MBR4045WT. The data sheet indicates an expected forward voltage of 0.5 V at 15 A. This gives 0.5 * 15 = 7.5 W power dissipation. In order to take advantage of fewer parts in the bill of materials, we can use the same heat sink as the 5.0 and 12.0 V supplies.

D10 reverse voltage will be the same as the switch peak voltage (twice HV input + transients), and the current is that required to discharge the parasitic inductances in the transformer. It will need to have at least 1000 PRV rating and current equal to the primary current. The HFA06TB120 FRED can handle both parameters. We choose a FRED for its fast response and its soft recovery characteristics.

The auxiliary supply has peak voltage of 81 V. The MBR1100 diode has 100 PRV and will handle 1 A forward current. The 1 A current rating gives plenty of margin.

Inductor Designs

The value of L1 will be determined by the desired ripple current. This output will not need very fast response, since the load is almost constant. We can choose 10% ripple current for this voltage at the worst case of highest voltage. In the case of an input voltage of 20 V, we have 10 V change across the inductor both while it charges and while it discharges. This means that the peak current will be exactly twice the average current. We can use the inductor equation to figure out the inductance that we need.

$$V = L \, di/dt \tag{11-11}$$

$$10 = L \, (2 \text{ mA}/5 \text{ μs}) \tag{11-12}$$

$$L1 = 10 * 0.000005/0.002 = 25 \text{ mH}$$

This choke has high inductance but low current, so a ferrite pot core or toroid will provide the required inductance and adequate magnetic shielding. An FT50 Mix 77 toroid with 151 turns of #28 wire will be our starting point. We choose #28 wire more for mechanical strength than for current capacity.

We use the same method to determine the values of L2 through L5: These supplies will need a higher ripple factor to enhance transient response. We will use 20% ripple factor for these voltages.

$$L2 = 12.0 * (0.000005/2.8) = 21.5 \text{ uH}$$

$$L3 = 3.3 * (0.000005/2.8) = 5.9 \text{ uH}$$

$$L4 = 12.0 * (0.000005/0.1) = 600 \text{ uH}$$

$$L5 = 5.0 * (0.000005/3.6) = 6.9 \text{ uH}$$

L2, L3, and L5 have high current, so Mix 26 toroid cores will provide the required inductance and magnetic shielding without saturating. L3 and L5 are close enough in value that we can use the same inductor for both, to have reduced bill of material costs. We will start with a T106-26 core for these inductors. A_1 for this core is 900 μH/100 turns. First, we calculate the number of turns required:

$$N = 100 \, (L/A_1)^{1/2} = 100 * (6.9/900)^{1/2} = 100 * (0.00767)^{1/2} = 8 \text{ turns}, \tag{11-13}$$

which gives a starting value of 5.8 μH for L3 and L5. Nine turns will give a value of 7.3 μH. A large DC bias current will decrease the inductance, so 7.3 μH is a good compromise for both inductors. Number 12 wire will give 40 C° temperature rise at 18 A for L5.

$$N = 100 \, (L/A_1)^{1/2} = 100 * (21.5/900)^{1/2} = 100 * (0.0239)^{1/2} = 16 \text{ turns}. \tag{11-14}$$

This gives an actual value of 23 μH for L2. Number 14 wire will give slightly more than 40 C° temperature rise, so we will want to ensure that the core loss does not overheat the inductor.

The value of L4 is large enough that we will need to consider using a ferrite core. An FT-50 Mix 77 toroid core will provide enough inductance.

$$N = 1000 \, (L/A_1)^{1/2} = 1000 * (0.600/1100)^{1/2}$$
$$= 1000 * (0.000545)^{1/2} = 23 \text{ turns.} \qquad (11\text{-}15)$$

Number 28 wire will be more than adequate for the 500 mA maximum current for the -12.0 V supply.

We need to verify the temperature rise and flux density for each of the inductors.

$$B = (E * t * 10^8)/(2 * A * N) = (L * \Delta I * 10^8)/(2 * A * N) \qquad (11\text{-}16)$$

$$\text{L1 } B = (25 \text{ mH} * 2 \text{ mA} * 10^8)/(2 * 0.133 * 151) = 125 \text{ G}$$

$$\text{L2 } B = (23 \, \mu\text{H} * 2.8 * 10^8)/(2 * 0.659 * 16) = 305 \text{ G}$$

$$\text{L3 } B = (7.3 \, \mu\text{H} * 2.8 * 10^8)/(2 * 0.659 * 9) = 172 \text{ G}$$

$$\text{L4 } B = (600 \, \mu\text{H} * 0.1 * 10^8)/(2 * 0.133 * 23) = 980 \text{ G}$$

$$\text{L5 } B = (7.3 \, \mu\text{H} * 3.6 * 10^8)/(2 * 0.659 * 9) = 221 \text{ G}$$

These values will allow us to calculate the temperature rise due to AC flux. We read the power density for the expected AC flux density from the graph for each material at 100 kHz.:

$$\text{L1 } P = (2 \text{ mW/cm}^3) * 0.401 \text{ cm}^3 = 0.8 \text{ mW}$$

$$\text{L2 } P = (400 \text{ mW/cm}^3) * 4.28 \text{ cm}^3 = 1.7 \text{ W}$$

$$\text{L3 } P = (150 \text{ mW/cm}^3) * 4.28 \text{ cm}^3 = 0.64 \text{ W}$$

$$\text{L4 } P = (300 \text{ mW/cm}^3) * 0.401 \text{ cm}^3 = 0.12 \text{ W}$$

$$\text{L5 } P = (100 \text{ mW/cm}^3) * 4.28 \text{ cm}^3 = 0.43 \text{ W}$$

We can use the power to approximate temperature rise:

$$\Delta T = (\text{Power/Surface Area})^{0.833} \qquad (11\text{-}17)$$

$$\text{L1 } \Delta T = (0.8/4.7)^{0.833} = 0.22 \text{ C}°$$

$$\text{L2 } \Delta T = (1700/22.6)^{0.833} = 37 \text{ C}°$$

$$\text{L3 } \Delta T = (640/22.6)^{0.833} = 16 \text{ C}°$$

$$L4 \ \Delta T = (120/4.7)^{0.833} = 15 \ C°$$

$$L5 \ \Delta T = (430/22.6)^{0.833} = 12 \ C°$$

We can see that L2 is likely to get too hot as currently designed. We will need to reduce the AC flux in the core. We will also need to increase the wire size to #12 to reduce the heat generated by the copper of the coil. We saw in Chapter 8 that decreasing the ripple current and increasing the inductance will have the largest effect because the number of turns will increase. We can also reduce the loss by using a core with a lower A_L value. Here are the new calculations for a T130 core with 1.4 A of ripple current:

$$N = 100 \ (L/A_l)^{1/2} = 100 * (43/785)^{1/2} = 100 * (0.0548)^{1/2} = 24 \text{ turns} \quad (11\text{-}18)$$

$$L2 = 785 \ (N^2/10,000) = 45 \ \mu H$$

$$L2 \ B = (45 \ \mu H * 1.4 * 10^8)/(2 * 0.698 * 24) = 188 \ G$$

$$L2 \ P = (180 \text{ mW/cm}^3) * 5.78 \text{ cm}^3 = 1040 \text{ mW}$$

$$L2 \ \Delta T = (1040/29.3)^{0.833} = 20 \ C°$$

We need to verify that the inductors will not saturate. The formula for magnetizing force is:

$$H = (0.4 * \pi * N * I) / 1 \quad (11\text{-}19)$$

where 1 is the magnetic path length.

$$L1 \ H = (0.4 * \pi * 151 * 0.02)/3.02 = 1.26 \text{ Oe}$$

$$L2 \ H = (0.4 * \pi * 24 * 14)/8.28 = 51 \text{ Oe}$$

$$L3 \ H = (0.4 * \pi * 9 * 14)/6.49 = 24 \text{ Oe}$$

$$L4 \ H = (0.4 * \pi * 23 * 0.5)/3.02 = 4.8 \text{ Oe}$$

$$L5 \ H = (0.4 * \pi * 9 * 18)/6.49 = 31 \text{ Oe}$$

We find that L1 is close to saturation and L4 is beyond the knee of the B-H curve. We need to increase the path length to reduce the magnetizing force. An FT82 Mix 77 core will have only slightly higher A_L but almost double the path length. The new L1 core will have the same number of turns and inductance.

$$L1 \ H = (0.4 * \pi * 151 * 0.02)/5.26 = 0.72 \text{ Oe}$$

$$\text{L1 B} = (25 \text{ mH} * 2 \text{ mA} * 10^8)/(2 * 0.245 * 151) = 68 \text{ G}$$

$$\text{L1 P} = (1 \text{ mW/cm}^3) * 1.29 \text{ cm}^3 = 1.29 \text{ mW}$$

The FT82 core is still not large enough for L4. We can use an FT114 core, which will need fewer turns because A_L is higher.

$$N = 1000 \, (L/A_l)^{1/2} = 1000 * (0.600/1270)^{1/2} =$$
$$1000 * (0.000545)^{1/2} = 22 \text{ turns} \qquad (11\text{-}20)$$

$$\text{L4 H} = (0.4 * \pi * 22 * 0.5)/7.42 = 1.9 \text{ Oe}$$

$$\text{L4 B} = (600 \text{ } \mu\text{H} * 0.1 * 10^8)/(2 * 0.375 * 22) = 363 \text{ G}$$

$$\text{L4 P} = (30 \text{ mW/cm}^3) * 2.79 \text{ cm}^3 = 0.084 \text{ W}$$

These cores will have minimum temperature rise because of the significantly lower power density. Note that the magnetizing force for L4 still places it close to the saturation point. This is likely to reduce the inductance at the limits of DC current. Laboratory testing may show that a different core will be required.

Capacitor Designs

The target internal temperature of the supply is quite high at 60°C, so we will need very high temperature-capable electrolytic capacitors. The CDE series 300 is rated for 2000 hours at 125°C. Derating to 60°C maximum will give 100,000 hours of life at 0.6 A rated ripple current.

C3 must produce less than 120 mV with 1.4 A ripple. The ripple rating needs to be $1.4/0.6 = 2.3$ A and ESR must be less than $(0.667 * 0.12)/1.4 = 57$ mΩ. The chart for ESR shows that the value is the same at 20 kHz and 100 kHz for the Type 300 capacitor. Assigning 33% of ripple to the capacitor requires $1/(2 * \pi * 100 \text{ kHz} * 0.028) = 57$ μF. The 1800 μF 16 V capacitor is the smallest that will satisfy our ESR requirement with 55.0 mΩ and 2.76 A ripple rating.

We repeat the calculations for C2, C4, and C5.

C2:

$$\text{Ripple rating} = 3.6 \text{ A}/0.6 = 6 \text{ A}$$

$$\text{ESR} = (0.667 * 0.05)/3.6 = 0.009$$

$$X_C = 0.005$$

$$C = 1/(2 * \pi * 100 \text{ kHz} * 0.005) = 320 \text{ } \mu\text{F}$$

$$\text{Actual } C = 18{,}000 \text{ } \mu\text{F}, 6.3 \text{ V}, 10.9 \text{ m}\Omega, 9.45 \text{ A ripple}$$

CDE 301R183U6R3JL2

C4:

$$\text{Ripple rating} = 2.8 \text{ A}/0.6 = 4.7 \text{ A}$$

$$\text{ESR} = (0.667 * 0.05)/2.8 = 0.012$$

$$X_C = 0.006$$

$$C = 1/(2 * \pi * 100 \text{ kHz} * 0.006) = 265 \text{ } \mu\text{F}$$

$$\text{Actual } C = 12{,}000 \text{ } \mu\text{F}, 6.3 \text{ V}, 15.3 \text{ m}\Omega, 8.27 \text{ A ripple}$$

CDE 301R123U6R3GS2

C5:

$$\text{Ripple rating} = 0.1 \text{ A}/0.6 = 0.17 \text{ A}$$

$$\text{ESR} = (0.667 * 0.12)/0.1 = 0.14$$

$$X_C = 0.07$$

$$C = 1/(2 * \pi * 100 \text{ kHz} * 0.07) = 23 \text{ } \mu\text{F}$$

$$\text{Actual } C = 820 \text{ } \mu\text{F}, 16 \text{ V}, 85 \text{ m}\Omega, 1.8 \text{ A ripple}$$

CDE 301R821M016EG2

Notice that C2 and C4 use significantly more capacitance in order to obtain a small enough ESR and enough ripple current rating. These capacitors are about 60 times larger than the required capacitance in order to meet all of the requirements.

Transformer Design

We will want to choose an E-type core for maximum heat dissipation. Table 4 of the Magnetics Design Application Notes lists cores appropriate to various frequencies and power levels. For 100 kHz and the 200+ W power level, the EC41 core is recommended.

The next step is to calculate the turns ratios required for the transformer. This design places difficult constraints on the operation of the DC–DC converter because of the wide input voltage range. In order to give the widest range of operation, we need to design the system to use the full range of pulse width. This means that we will set the pulse width to 50% at 100 V input. The system will lose control just as we reach the lowest voltage during a power failure.

The voltage across the inductor for a forward converter is equal to the output voltage for 50% duty cycle. We will design the auxiliary power supply to provide maximum 20 mA. The transformer winding voltage will be double the output voltage. We must supply additional voltage to overcome the voltage drop across the rectifier. We will require 20.3 V (includes 0.3 V for the diode) for the auxiliary winding. This gives a turns ratio 100/20.3 = 4.93. The voltage drop for a Schottky diode at 18 A is approximately 0.7 V. This means we need a transformer winding of 10.7 V at the lowest input voltage for the 5 V supply. The +12 V supply will need a minimum voltage of 25.2 V because we must use an ultra-fast diode instead of a Schottky diode. The 3.3 V supply will require a 7.1 V secondary winding. The 5 V turns ratio is thus 100/10.7 = 9.35, the 12 V turns ratio is 100/25.2 = 3.97, and the 3.3 V turns ratio is 100/7.1 = 14.1.

We can start our analysis by choosing 20 turns for the primary. We have to ensure that the flux density will not overheat the core for the highest input voltage.

$$B = E/(4 * A * N * F * 10^{-8}) \tag{11-21}$$

E = RMS voltage (P–P/2 for a square wave)
A = core magnetic area in cm^2 (from data sheet)
N = Number of primary turns
F = frequency in Hz

For our core,

$$B = 195/(4 * 1.24 * 20 * 1e^5 * 1e^{-8}) = 1965 \text{ G}$$

Type R material from Magnetics provides a reasonable loss at 2000 G and 100 kHz. From the material charts, we see that the core loss is 400 mW/cm^3.

$$P_{LOSS} = 400 * 10.9 = 4.36 \text{ W}$$

This is a fairly large power loss and is likely to cause a high internal temperature. Doubling the number of turns will decrease the flux to 982 G and decrease the core loss to 70 mW/cm³. It will also allow us to have a better match between the various supplies by having more accurate turns ratios. The power loss drops to just 763 mW.

This allows us to set the windings:

Primary	40T
Reset winding	40T
Aux. winding	8T
3.3 V winding	3T
5 V winding	4T
12 V winding	10T

The primary current is 2.8 A, so #16 wire will handle the current for the primary winding and the reset winding. The bobbin for the EC41 core has a winding length of 0.965 in., so we will have 18 turns per layer. The primary and reset windings have a common connection that is phased 180 degrees, so it is possible to wind these two windings as essentially one 80-turn winding with a connection in the middle. This will require five layers with a large amount of space left on the fifth level. The auxilliary winding can be wound using the space left on the fifth layer, using a convenient size of wire. Number 22 wire will give enough space to meet safety separation and fill the space on that layer.

The windings for the 3.3 V, 5.0 V, and 12.0 V high current windings should be wound using a 0.8-in. wide copper strip that is 20 mils thick. Number 22 wire is adequate for the −12.0 V winding.

The next step is to produce a prototype and take it to the lab for testing!

Index

Universal input, 54, 115, 121, 129, 131, 144, 160, 196, 224, 302

V

Voltage doubler, 49, 53, 67, 70, 160
Voltage mode PWM controller, 23, 25, 26, 33, 40, 63, 152, 153

W

Wire table, 239

Z

Zener diode, 46, 69, 115, 127, 147, 164, 172, 215, 216, 307

www.ingramcontent.com/pod-product-compliance
Lightning Source LLC
Chambersburg PA
CBHW080919220326
41598CB00034B/5619

* 9 7 8 0 7 5 0 6 7 4 4 5 4 *